MW00714381

THE CONFRONTATION: BLACK POWER, ANTI-SEMITISM, AND THE MYTH OF INTEGRATION

The Confrontation: Black Power, Anti-Semitism, and the Myth of Integration

by Max Geltman

PRENTICE-HALL, INC.
Englewood Cliffs, N.J.

For Rae and Toby

The Confrontation: Black Power, Anti-Semitism, and The Myth of Integration

by Max Geltman

Library of Congress Catalog Card Number: 74-105862
Printed in the United States of America • *T*
ISBN-0-13-167528-1

Prentice-Hall International, Inc., London
Prentice-Hall of Australia, Pty. Ltd., Sydney
Prentice-Hall of Canada, Ltd., Toronto
Prentice-Hall of India Private Ltd., New Delhi
Prentice-Hall of Japan, Inc., Tokyo

PREFACE

I offer this book without footnotes (with necessary references embodied in the text). This is not another sociologist's handbook, but my statement of the observed and observable facts. The sociologist has for too long intruded his "expertise" on a situation which he has done little to alleviate—much to aggravate. Ralph Ellison perhaps put it best when he said ". . . we depend upon outsiders—mainly sociologists—to interpret our lives. [Too often] we lose the capacity for abstracting and enlarging life. Instead we ask, 'How do we fit into the sociological terminology? Gunnar Myrdal said this experience means thus and so . . .' Well, whenever I hear a Negro intellectual describing Negro life and personality with a catalogue of negative definitions, my first question is, 'How did you escape, is it that you were born exceptional and superior?' "

MAX GELTMAN

Contents

Part 2 The Negro Family Problem

Part 3 The Contemporary Scene

Part 4 The Confrontation

THE CONFRONTATION: BLACK POWER, ANTI-SEMITISM, AND THE MYTH OF INTEGRATION

introduction

The Negro-Jewish Confrontation

THE NEGRO-JEWISH CONFRONTATION

Introduction

In March, 1935, following the first massive riots in Harlem, Adam Clayton Powell, Jr. said that "the Jews own New York, the Irish run it, but the Negroes enjoy it." There is enough of a half-truth here to make it an interesting observation. But the facts are the Negroes had very little to "enjoy" in that depression year; the Jews "owned" nothing but a lot of bad debts, and La Guardia (not O'Brien) was Mayor of New York.

In the years that followed, Harlem saw other riots as the confrontation heightened. But it wasn't until the sixties that overt Negro-Jewish antagonisms began to make themselves manifest, not only in New York but in all the Harlems of the land. By the time 1969 came around anti-Semitic sentiments had become so prevalent that it became fashionable to speak contemptuously of the Jews in the highest echelons of established cultural power—including the Metropolitan Museum of Art.

On January 19, 1969, the museum opened a special show titled "Harlem on My Mind." The show was largely a pictorial representation (with sound and blinking lights) of the Negro community from 1900 to 1968. It was picketed by black militants who said it did not represent in any way the creative aspects of Harlem life, while more moderate Negro leaders like Kenneth Clark condemned the show outright and demanded it be closed down. The cluttered and often badly identified photographs of former Negro leaders, unprepossessing shots of Harlem tenements, and the noise and confusion disturbed the critics, causing them to condemn the show on aesthetic grounds. Jews were outraged by an introduction to the museum's catalogue which said, among other things, that "our contempt for the Jew makes us feel more completely American in sharing a national prejudice." Whether that statement demonstrated that the whole nation was anti-Semitic is not the point here. But that a distinguished museum would print such rot in one of its official publications stunned the city. Unfortunately, the unhappy and childish bit of anti-Jewish bias expressed in the Introduction *was* written by a child; that is, it was written by a Negro schoolgirl, age 16, and the high potentates at the museum found it a fitting expression of Negro sentiment in the Harlem community.

No sooner did the museum flap die down than Station WBAI in New York broadcast a "poem" read by Leslie Campbell, a teacher at Junior High School 271 in Ocean Hill-Brownsville. The epic which he attributed to an anonymous 15-year-old student (where did this student get such vile notions?) said, in part: "Hey, Jewboy, with the yarmulka on your head/You paleface Jew boy, I wish you were dead." There are some 38 additional lines of similar poetic insights in this effusion, replete with Israel-hatred and other poisonous nonsense, all of it dedicated to Albert Shanker, head of the New York Teachers Union. And, as if that were not enough, a few days later the same station broadcast a statement from a black student at New York University to the effect that "Hitler did not make enough lampshades out of you Jews."

There were those in the metropolis who found a variety of defenses for these atrocious statements; the museum carried on in its sublime way; the radio station garnered a bit of needed publicity; and the teacher and student who spoke their Jew-hatred in public went on teaching and studying at the public expense—as if it were the most normal thing to say and do in New York City in 1969. In fact, Roy Innis, head of the Congress of Racial Equality (CORE), declared in February of that year that "a black leader would be crazy to publicly repudiate anti-Semitism." And on May 27, at a meeting sponsored by Independent CORE (a breakaway movement from the parent organization) at Junior High School 111 in Brooklyn, a spokesman opened the discussion with these words: "If a black teacher says that Hitler killed three million Jews and he's told six million, then Hitler was a better man."

Incidents like the ones just cited were not confined to New York City and its environs. In Detroit things got so bad that the president of Wayne State University declared that the student newspaper, edited by an off-campus black militant, was publishing material "disturbingly reminiscent of Hitler Germany." Similar outrages were reported from other cities, and on August 30, 1969, the *Black Panther* (official organ of the Black Panther party, published in San Francisco) denounced "Zionism" as "kosher nationalism," which it said equaled "fascism." The same article also spoke of "Zionist fascist pigs."

At just about this time, Floyd McKissick, who had spoken loudest for Black Capitalism during the Nixon campaign, was bypassed when it came to a federal appointment. (James Farmer, who didn't say a good word for Nixon or Agnew all through the campaign, received the important assignment as special administrative assistant for urban and suburban affairs.) Finally, McKissick (perhaps because he has little faith in capitalism, white or black—regarding the latter as a means for disintegrating the former), for the third time, declared that he was ready to lead his disciples from the Egypt of Bedford-Stuyvesant in Brooklyn and deliver them to North or South Carolina to establish there a sort of black kibbutz.

McKissick's intention is serious, and should not be ridiculed in advance. But at a time when other black spokesmen—especially of the instant, media-made variety—are clamoring for "local control" of the black communities from the school to the firehouse, Mr. McKissick's plan seemed doomed before it got started. His quest to emulate the kibbutz experience of the Israelis is commendable, if foolish. Foolish, because alien social experiments are not easily borrowed, and if borrowed are usually doomed to failure. After all, the Jews built *kibbutzim* to prove something to themselves, namely, that they could, if given an opportunity, work the land. The Negro in America does not have to prove anything of the kind.

False analogies will not help the black man receive his fair share in the nation. Telling Negroes how Jews "made it" in this country in the face of adversity stirs even the most moderate of Negro leaders (Bayard Rustin, for example) to outrage. Not that Rustin and others are justified in their wrath; indeed, I think they are often quite wrong. That the outrage exists is what counts. Among less sophisticated blacks, it is expressed in the "Jew bastard" lingo of the gutter. Those on the intellectual level of James Baldwin rationalize these outbursts on the grounds that all communities need objects for their hatred: "Georgia has its Negroes, Harlem has the Jew."

The forced family equation is another case in point. The problem of the Negro family haunts the conscience of the nation. Yet the advice usually offered—gratuitously in most cases—will not help to solve the problem. It must ultimately be solved within the greater Negro community itself. The solution must be based on Negro folkways, not on those of other ethnic groups, some of which have familial patterns based on centuries-old traditions. In the case of the Jews, thousands of years of persecution have helped create not only a unique kind of family coherence, but have imbued them with a spirit of survival that is perhaps their most important achievement. Mark Twain once observed that if the Jews have a genius for anything, it is the genius for survival.

American society has never been static, and no group has for long been denied the fruits of its own efforts. Success has, however, been slow for some and slower for others. It has never been easy for any. "We are all immigrants" is the grand American cliché. However, most of us came here voluntarily (more or less); the Negro did not. If, today, he likes to call himself an "Afro-American" or a "black man," underscoring his racial background rather than his American identity, his choice is understandable, if regrettable. He will have to learn for himself that Africa holds no promises for him, as Richard Wright and others have learned to their sorrow.

Despite all this support there has grown up in the Negro communities of the nation—especially in the urban centers from Harlem to Watts—a hatred of Jews that baffles the liberal community as a whole and the Jewish community in particular.

It is a part of the aim of this book to attempt to explain this unusual phenomenon in terms of the dual image the *ordinary* Negro has of the *ordinary* Jew, especially in the large "ghettos" where Negroes and Jews come into frequent contact as customer and merchant, as tenant and landlord, and sometimes simply as black and white.

This dual identity of the Jew is not a new one. It was visible in Germany before the holocaust. In one of Hitler's early speeches we hear of the Jew—as exploiter and benefactor—in these words: "Moses Kohn on the one side encourages his associates to refuse the workers' demands, while his brother Isaac in the factory, invites the masses to strike." This contradictory image of the Jew which helped confuse the ordinary German 40 years ago, is to a large extent responsible for the distorted image of him that is held by the Negro man in the street. The Negro is baffled by the Jew who, on the one hand offers him considerable aid, and with the other seems to exploit him as the owner of the delicatessen around the corner or the pawnbroker to whom, as the Negro poet Langston Hughes wrote, he brings his old clothes.

In almost all instances of anti-Semitism it begins with the leaders, the intellectuals, for *all* leaders are intellectuals in the sense that they formulate ideas, even if they sometimes express themselves in the hip lingo of black power or the scalding strophes of the Negro poet who demands that his audience "rip the Jew-woman's belly open." The Negro anti-Semite is no innovator here. White intellectuals—from Voltaire, who despised Jews, to Louis-Ferdinand Céline, who offered them up as "bagatelles for a massacre"—have stoked the fires of hatred for Jews during hundreds of years.

Yet when the cacophony of hate reaches into the Academy and the Museum; when the Hitler-lampshades imagery is revived—as it was in an attack on a leader of the American Jewish Congress by a CORE official in Mount Vernon; when that imagery is repeated four years later in a radio broadcast; when the six million burned and gassed in Hitler's ovens are mocked at a speakout by a couple of black intellectuals; when Negro children start talking of "pale-faced Jewboys"; when contempt mounts and threats follow and Jews begin to set up their own self-defense groups; when riots in Watts, Cleveland, Cincinnati, Newark, Washington, Rochester, Detroit, Harlem, and elsewhere imperil —in most cases—Jewish storekeepers, there is cause for alarm. And no amount of self-serving surveys will obliterate the matter, nor will they help "solve" the problems as some statisticians hope, on the premise that what is not recorded will not be noticed. The confrontation is genuine and cruel. And some of the best among the Negro leadership have recognized the fact.

The confrontation did not come upon us out of the blue—without sign, without warning, without those semaphores that could be read. Most of us chose to ignore them. With the advent of the massive riots of 1967, the visible signs of the conflict became a clear and present danger to the Jewish minority, which was sure it had done more to help the downtrodden Negro than had any other group in the land. And so it had. But this, instead of making the black man grateful, only helped to "increase his rage," as James Baldwin noted.

Then, on May 19, 1969, events seemed to have come full circle with the outbreak of another riot in Newark, New Jersey. (An analysis, in Chapter One, of some of the riot-torn centers in 1967 shows that the cities designated as "best for Negroes" turned out to be best for rioting, too. Among them, Newark was the first city in which an "unanticipated" riot occurred.) The new riot, the result of the shooting of a black youth by a Negro policeman, again saw shops looted, with white merchants—very likely LeRoi Jones' small "joosh" shopkeepers—being dragged out of their shops and beaten.

In all, the summer of 1969 was not as "hot" as the previous two summers had been. Only 95 cities felt the lash of black rage as against 125 in 1968 and 138 the year earlier. *Time* magazine, commenting on these outbursts, notes that the most serious aspect of the recent riots was the fact that following "the disorders that shook Washington after the murder of Martin Luther King, Jr., ghetto rioters learned that downtown stores and prosperous neighborhoods can be made as vulnerable as their own. For many citizens that legacy is far more troubling," *Time* concludes, "than all the rhetoric and social studies with which official Washington has documented the spread of crime." And in December, 1969, Adam Clayton Powell issued a warning that "Downtown, where the big department stores are located, is a much better target for black anger than is the ghetto."

One thing the riots demonstrated was that it was not the poor who rioted (as the sociologists had taught and the politicians had repeated). For not only is poverty no crime, it does not even lead to crime, else we'd all be criminals since most of us have been poor at one time or another—and some of us still are. Nor will it help the Negro to tell him the solution to his problems lies in "abolishing poverty." Poverty being a relative condition, it can never be done away with entirely. What is poverty to Nelson Rockefeller would be fabulous wealth to most of us; and Mrs. Aristotle Onassis (Jacqueline Kennedy) would feel absolutely impoverished if she were compelled to cut her annual wardrobe allowance down to a measly hundred thousand dollars. That does not mean that one should not make every effort to solve the problem of hunger and malnutrition, but even here one finds the subject exploited in selective statistics, political maneuvering, and false promises. After all, some of the richest people in America are malnourished, from either dieting too much or too little. But hunger must be rooted out of the land, for it just has no place in America. The demagogues of both races who think they have found in the "poor" (the fashionable euphemism for Negro) a way station to power, in the streets or in the legislatures, may nevertheless be in for some disappointments. (In the last elections for Councilmen in New York City, one Jesse Gray, the most abrasive—and largely self-appointed—spokesman against black poverty in Harlem, was roundly defeated by a moderate Negro who promised less with the hope that he could deliver more.)

Meanwhile, black militant leadership had shifted from the Stokely Carmichaels and the Rap Browns, whose voices had been silenced by self-imposed exile or because of indictments in the courts. For a while it appeared that men like Eldridge Cleaver would spearhead the drive for a new black insurgency, with the Black Panther party in the forefront. Troubles with the law soon removed Cleaver from the scene after he jumped bail, going first to Cuba (where he was unhappy), finally lighting in Algeria. Here he has joined Carmichael, who operates from a neighboring North African country, in a campaign to help liberate Palestine from the Jews.

As 1969 was drawing to a close the Black Panthers came under heavy fire growing out of confrontations with police, resulting in the shooting of more than a score of Panthers (and police), and the imprisonment of almost the entire leadership. And among the first to come to the aid of the beleaguered Panthers (after the Chicago shootout in which two Black Panthers were shot dead by the police) was former Supreme Court Justice Arthur Goldberg, and his organization, the American Jewish Committee, as well as the American Jewish Congress.

The Goldberg involvement may prove risky at best. Certainly it cannot help defuse the situation as far as the police and black militants are concerned. Neither the Committee nor the Congress has any special expertise here. Should the Goldberg committee find for the police, the hatred from the black community will be intensified tenfold. Should it find for the Panthers, it will probably help fortify the latent Jew-hatred in the bosoms of millions of American workingmen whose sense of exasperation at the number of Jewish-identified names in college eruptions and street demonstrations will almost certainly be exploited by the next anti-Semitic Populist who comes along—a tradition that has deep roots here.

We can be sure that the black rage that prompted a Panther poet to write "Jew land, On a summer afternoon/ Really, Couldn't kill the Jews too soon," will not be assuaged by Mr. Goldberg's intrusions in an area where his ministrations can only be misunderstood, despite the excellent motives that no doubt prompted his entry into the dismal situation. Finally, on the last day of the year, Eldridge Cleaver, head of all the Panthers, even in absentia, issued a manifesto calling on his followers in America to "support the armed struggle of the Palestinian people against the Israeli watchdogs of imperialism," adding, "Zionists, wherever they may be, are our enemies."

The black intellectuals, those the nation had long known, were now strangely silent. Ralph Ellison did not utter a word after his altercation with the editor of *Commentary*, a journal sponsored by the American Jewish Committee. (The significance of this will be found in a later chapter.) James Baldwin was cultivating the muses somewhere abroad, and LeRoi Jones was engaged in producing a kind of hate-honky drama that gave vicarious enjoyment to the middle-class white offspring of well-to-do parents in suburbia, when he wasn't involved in the politics of Newark, (a city that was to receive national attention at the end of 1969 when almost all public officials were indicted on federal warrants on a variety of counts of corruption). But other intellectuals, men like Julius Lester, were getting a kick out of having their works publicly advertised in a Jew-baiting and honky-hating kind of lingo, like the ad in the *Times* which teased the reader by calling his attention to the fact that "A Tampa Lynch Mob and the New York Rabbis are Both Dying to Get this Man."

With Carmichael gone and Cleaver on the lam, with others silent or imprisoned, new voices and faces were being seen and heard. James Forman re-emerged demanding a half billion dollars in "reparations" from church and synagogue for past wrongs—and future deliverance; the while Roy Innis was demanding no less than six (6) billions in reparations from the business community of the nation. But among the more successful was George A. Wiley who heads up the National Political Welfare Rights Organization. Wiley believes that there is a massive political potential in the unemployed blacks on relief.

With President Nixon's proposal of a federal base of $1600 for all poor families of four—a dream-come-true for Daniel Patrick Moynihan, who had tried vainly to have two previous administrations adopt the idea—or what amounts to a guaranteed annual wage, a national conference on hunger was called early in December, 1969. Under Moynihan's guidance and supervised by the activist French nutritionist Dr. Jean Mayer, it was the first such conference ever held. And while the more militant delegates were not satisfied with the results (they wanted $5400), still it was a beginning. So that even if Mr. Lester was howling, "Bang! Bang! Mr. Moynihan," others saw many virtues in a commitment to a work-oriented program which would permit families on relief to keep added wage earnings without incurring penalties as had been the case before the President's proposed program. The plan was also designed to help keep poor blacks down South, to deter them from seeking the fleshpots of easy and relatively huge relief allotments in the richer North. Also—and most important—it was hoped that it would help keep the (missing) Negro father in the home.

1969 was also the year of the Conferences. Conferences on hunger and crime and manpower resources and urban decay and pollution and whatnot became the order of the day. But the First Black Economic Conference introduced a new wrinkle to Negro demands with Professor Robert Browne of Fairlegh Dickinson University advising that "Racketeering, and the numbers, if they are to continue, must be put into the hands of the black community." After this, it could be truly stated that all the black man's panaceas were now behind him.

The present book is the outgrowth of an essay written four years ago and published in *National Review* under the title, "The Negro-Jewish Confrontation." It was my hope that by making clear the "dual identity" in which the Negro sees the Jew—ambivalently, as exploiter and benefactor—it would help both to limit the area of conflict and to see it in its proper dimensions.

But the forces of an extremist integrationism ruled the temper of the times, as do today the forces of another kind of extremism: the advocacy of total separatism—apartheid—that has attracted the black revolutionaries in the colleges and in the streets. A good deal of this is undoubtedly due to the disappointments engendered by the myth, spearheaded by the Supreme Court decision of 1954, which argued that black children will learn better when sitting next to a white child than in an all-Negro school, regardless how excellent the curriculum or the teaching staff. (Recently Victor Solomon of CORE has gone South to test the thesis, by setting up all-black schools that will be not only separate and equal—but separate and better, he hopes.)

When Roy Wilkins spoke out bitterly against the tendency to self-segregation in the form of Black Studies departments and black dormitories in our universities, he had in mind the hurt implicit in a situation whereby he could see the ultimate harm such separatism would achieve. At a time when the nation needs qualified men to enter viable professions, when the nation needs men of any race in a variety of specialized skills, the Negro is being shunted off into an area of self-indulgence and a new colonialism, egged on by certain black demagogues who have borrowed an alien ideology that promises the Negro in America nothing but further frustration. But the white masochists and flunkies and Uncle Toms say," Let them have it. What the hell! It ain't going to do them no good, so what harm can it do us. All it will cost is a bit of money—and that we can always scrounge up from a frightened alumni body or a panicky state legislature." The point being that setting up Black Studies departments is not designed to help the black students, but it does help to bail supine college administrations out of a dilemma. Most responsible Negro educators are aware of this. Sir Arthur Lewis, professor of economics and international affairs at Princeton, sums it up when he says, "The road to the top is through higher education—not black studies." And Bayard Rustin adds, "What the hell is a Negro going to do with soul courses?"

They won't get him a job with IBM; they won't help build a bridge over a raging river; and they will not provide another physician in a slum neighborhood that could use all the doctors it can find. But they will make for a spurious kind of Black Power. Not the power of self-reliance and self-achievement, but the power that comes out of the barrel of a gun, the slogan of the Black Panthers. Yet, for all this, very few Negroes want any part of separatism or Black Pantherism. Most Negroes reject with loathing the epithet "pig," which Eldridge Cleaver pinned on the late Robert F. Kennedy, and they feel shame when the co-founder of the Panthers, Bobby Seale, calls Supreme Court Justice Thurgood Marshall "a bootlicker, a nigger pig, a Tonto and a punk."

But certain white intellectuals love it. And among them can be found many Jewish intellectuals who, ashamed of the faith they have long abandoned, alienated from the nation that has made them rich, revel in self-hatred by underscoring their unconscious (but real) contempt for the Negro who can make it—and often does—by wanting their "nigger" to be a rapist, a felon, a hatemonger, a racist. That is what Cleaver is to them. It is these white and emasculated and shameless and despoiled and renegade Jews—along with many of their ultra-liberal and WASPish friends—who have helped exacerbate Negro-Jewish *and* Negro-white relations in America. They are the ones who swoon at the literary graces in Panther literature such as the dirty little rhyme that says, "The Jews have stolen our bread. Their filthy women have tricked our men into bed."

It may well be as the president of the Synagogue Council put it when he said that many highly-placed whites are serving as "coat-holders so they can enjoy the Negro and Jew in conflict." But it is more than that. It is a conflict that threatens to tear the nation apart. Some on the far left would, of course, welcome that. Others—acting out of mistaken goodwill or an overweening sense of righteousness which they sometimes translate as "social justice"—are only serving as intruders in the choking dust of black-white relations in America. Says Harold Cruse in *The Crisis of the Negro Intellectual:* "In fact, the main job of researching and interpreting the American Negro has been taken over by the Jewish intelligentsia to the extent where it is practically impossible for the Negro to deal with the Anglo-Saxon majority in this country unless he first comes to the Jews to get his 'instructions.'" This may be exaggerated, but there is much truth in the statement.

That is why this book calls for a disengagement—the one volatile minority from the other. In a word, the Jew has trouble enough right now. It is time he looks back at the source of his own pride, and not volunteer advice and expertise—even where he has it—to those who would rather, right now, try to make it on their own.

It is relatively easy for the Jew to alter the altruistic side of the image he presents to the Negro community. All he has to do is to stop "giving." As the numerous citations in the book indicate, it is not often appreciated. It is more difficult, however, to rid himself of the "exploitative" side, although on this count he is certainly less guilty than charged. As landlord he is no more culpable than is his black counterpart. As merchant he charges no more than his gentile—black or white—competitor does, despite all the nasty propaganda heard from the most unusual sources. But as the onus is on the Jew, for whatever reasons, he will have to leave the black "ghettos" and to leave them in good faith.

It would not be the first time the Jew has left to make new beginnings. But he must not let himself be driven out. What he is being asked to provide is an accommodation, not a "right" due to others. The law (without which he is helpless) must provide protection for him until he is ready to depart—in peace.

For the Jews, history has been a series of arrivals, settlements, and departures. The shopkeeper need not go off to Israel if he doesn't want to. He can go to almost any other underpopulated state in the Union. He can retire or he can go into another line of business. His children have all the diplomas, and he still has the ability—much abused these days—of knowing how to make a living. He knows, too, that he does not go to others for aid, and wonders why should the Negro?

Today only this prospect makes sense for Negro and Jew alike. Relying on others' good faith seems a hollow mockery to members of the "remnant" left after the last great holocaust which the world watched in silence. The blind passions of unruly people—even a meager 5 percent—may be too much for the Jews in the United States. It may turn out that "the fire next time" will be for them, and they have had enough of fires.

The Negro wants to go it alone. The Jew has always gone it alone. This is better ground for understanding between them than the pious sentiments of a "common" minority identification when all each wants is to be taken separately for what he is.

part 1

The Events

Chapter One
THE 1967 RIOTS BEGIN

Early in the spring of 1967 a group from Brandeis University's Lemberg Center for the Study of Violence issued a report that seemed to offer a clue to the problem of urban violence in American cities. Its researches compared San Francisco, Cleveland, and Dayton, which had all experienced riots the year before, with Pittsburgh, Boston, and Akron, which had not. This comparison seemed revealing, but before the report could be properly digested by the press, a series of violent upheavals shook Boston, a city where riots were not expected. (We shall see later that much the same mistake was made in connection with New Haven and Detroit.)

Early in June the Roxbury section of Boston exploded in rioting and looting that presaged the long, hot summer the whole nation was in for. The first recorded victim of the rioting—of the Black Revolution, as it would later be called by certain incendiary leaders—was fifteen-year-old Elain Klein.

As often happens reality had upset the neat conclusions of the professors, who are, perhaps, less in a position than anybody else in the entire population to understand what is really going on and least likely to come up with any viable solutions to what is a major problem confronting the nation. If, as it has been argued, war is too serious a matter to be left to the generals, then society—and the life of us all—is too important to be left to the academicians.

This modest group from Brandeis announced that, "If we are able to help reduce violence in any way in the summer ahead, we will be gratified." Indeed, the whole nation would have been equally gratified.

Meanwhile, the weather was heating up as the summer solstice approached. By the end of June a series of riots had broken out in cities across the nation, with a fairly big one in Buffalo that lasted three days. That one might possibly have been halted earlier if the mayor of the city had not begged the rioting mob to "give him a chance." He promised to call off the police if the mob would behave as decent citizens. But the rioters only hooted defiance. "I will see to it that if you're not looting or burning, the cops will not be here," he cajoled. The mob went on a rampage that lasted nine hours, during which were committed some of the wildest outrages that occurred in that city.

By the time the summer had reached midpoint, certain aspects of the Negro-white conflict, never before clearly articulated, had begun to reveal themselves. Although it had been foreseen as early as 1958 that "an emerging Negroism would lead to growing anti-Semitism," this likelihood had not been widely recognized—at least publicly—at the time. That confrontation between white and black might lead to racial violence engulfing an entire nation had long been the "lunatic" prediction of certain racist anthropologists abroad but had been given little credence in this country. As early as the last years of the nineteenth century the Frenchman Vacher de Lapouge had warned that in the twentieth century nations and peoples would destroy one another because of "a quarter inch more or less of cephalic index." But that time had not yet arrived, or had it?

In early 1967 Professor Nathan Hare of Howard University announced that he detected something new in the struggle for Negro rights. According to him that struggle had taken on the complexion of a "racial war." And, with the approach of summer, the heat and durability of which had been predicted all winter by leading Negro spokesmen, including Reverend Martin Luther King, it was indeed possible to detect a change in the climate of opinion in this country. Frantic efforts to prevent the holocaust met with varying degrees of success in the cities. Antipoverty projects for special training and special jobs, for summer recreation and summer schools, for rat control and pest control, for more housing and medical centers, for whatever other panacea the mayors thought would capture the imagination of the "poor" were pushed through. Not knowing where the fire would strike this time, some politicians had reached the gnawing conclusion that it really did not matter what they did; they kept up a drumfire of good cheer and cooperated with whoever was willing to talk about ways to head off civil strife. Deep in their hearts they must have known that, despite all the bragging, bluffing, and catering to hoodlum elements, the likelihood of upheaval in the cities could not be wished away.

The first *big* unanticipated riot hit Newark like an exploding volcano in Mid-July. For one week an uncontrolled mob held sway over a third of the city, burning and looting like a cloud of locusts devouring everything in its path. By the time the fever had spent itself, 25 people were dead and millions of dollars worth of property had been destroyed or stolen. Firemen, rushing to put out fires, were shot at. The hospital where the wounded were carried was under siege, and inside there was rioting in the wards. So avid was the lust for destruction that even hospital furnishings were broken. The national guard was called out to help put down the "insurrection," as it came to be called by both Governor Richard Hughes and Stokely Carmichael. Finally, the riot tapered off. Blame could be placed in large measure at the doorstep of the so-called "militants," aided by lumpen elements of both the middle class and the unemployed and by a supine city administration that refused to act with firmness before the riot got out of hand.

What caused the riot? Certainly not the fact that a Negro taxi driver had been given a ticket by a white cop, nor that the city council had voted to erect a medical center in the heart of the so-called "ghetto," or that the Mayor had refused to appoint a Negro to the board of education. The cause was all these things together—and more. And there was outside interference of a sort.

While there was still peace in Newark and the city's people thought no more of riots than did those in other concerned communities, the Federal government, through the Office of Economic Opportunity, set up a local United Community Corporation that acted quite independently. It was amply staffed and its board of directors included one Willie Wright. This "outside" agency, it has been charged, did much to disturb the fairly cooperative relations in this community, which had two Negro city councilmen and a well-meaning liberal Democrat, Hugh Adonizzio as mayor. But neither the Mayor nor Police Commissioner Dominick Spina could cope with the Federally sponsored and Federally funded Community Corporation. (Adonizzio would later be indicted on a number of charges, including extortion and income-tax evasion. But even before that, many in the city had known of another type of outside influence, this one emanating from the underworld, that pervaded every aspect of official metropolitan life. But this influence was not essentially racial in origin, and contributed nothing to the riots, one way or the other. After all, the Mafia, too, is color-blind.)

The quality of Mr. Wright's influence and guidance can be gathered from the following remarks attributed to him by the New York *Post* on August 1, 1967: "This is what I am advocating around town: Get yourself a piece [gun] and put it in the bottom drawer or somewhere and have it fully loaded. Then if some joker cop or trooper breaks into your house, let him have it." Naturally enough, when all the shooting began and the fire-bombing was at its height and the police and national guardsmen were trying to restore order, some of Mr. Wright's disciples may have decided to "let them have it." And it would not be at all surprising if some of them acted out of a conviction that they were obeying a Federal mandate. After all, the Negro community knew that Willie was a Federal man—and that his agency was not accountable to Chief Spina. The fact is that months before the riots he had tried to warn some members in Congress of the buildup in Newark by the forces of Willie Wright, but no one had paid him any heed. (In a post mortem, at police headquarters, three months after the riots, officers looked at faces on a televised screening of the pre-riot attack on the Fourth Precinct Headquarters on July 13: "Face after face they identified as workers from the United Community Corporation, Newark's antipoverty agency" [*The New York Times*, October 14, 1967].) In fact, this aspect of the turbulent events in Newark has been pretty much ignored in the official investigations that Washington has engaged in.

The question of "outside influences" was raised by many people interested in the causes of the riots. Surprisingly, the Kerner commission's thousand-page report devotes less than two pages to this in Chapter 3. This part of the Report is called "Organized Activity," and concludes:

> On the basis of all the information collected the Commission concludes that the urban disorders of the summer were not caused by, or were they the consequence of, any organized plan or conspiracy.

However, the report adds:

> Militant organizations, local and national, and individual agitators, who repeatedly forecast and called for violence, were active in the spring and summer of 1967. We believe that

> they deliberately sought to encourage violence, and that they did have an effect in creating an atmosphere that contributed to the outbreak of the disorder.

The Newark riots proved contagious. In a few days—they had spread to such other New Jersey communities as Plainfield, Paterson, and Englewood. Of the turbulent Plainfield riots (they would later be cited by H. Rap Brown as the model for all future riots) *New York Times* reporter Paul Hoffman wrote, "Among the most enraged participants in the violence that has struck this affluent suburban city like a tornado, are teen-agers from its small Negro middle class."

Actually these "affluent" teen-agers were rioting against the cant of liberalistic leaders, black and white, who equate the Negro's condition with his lack of education, lack of money, lack of material possessions, in a word, his poverty. Liberals were challenged to see the riots in a new light. In the Dixie Hill section of Atlanta, with its tree-shaded white frame and stucco houses; in Waterloo, Iowa; in posh Nyack, New York; and in affluent Plainfield, Negro children of middle-class parents do not seek civil rights; they already have them. They do not seek "integration"; it does not "work." Even poor Negro youths do not want welfare handouts; welfare shames them as it apparently does not shame the social workers (for they are as much on "the welfare" as are any of the "cases" they "service"). And Negro teen-agers do not want government-made projects to create work for them.

They want to find their own jobs, start their own businesses, and work out their own destinies as best they can. They are seeking self-awareness. They are seeking their own kind of status, their own kind of dignity, their own kind of pride. Since no one in all America seems to understand them, much less to help them; they must turn to the soul-destroying lingo of uninformed young men who, tired of being told that poverty is their crime, turn to the alien and destructive ideologies of men like Mao Tse-tung, Frantz Fanon, and Che Guevara.

The great dream of total integration in a culturally pluralistic society has turned out to be—as might have been expected—nothing more than a dream for the Negro.

Chapter Two
THE NEWARK CONFERENCE ON BLACK POWER

Newark was soon forgotten, as other riots gripped the nation, but not before it went through with the scheduled Conference on Black Power, held July 20-23 at the Cathedral House headquarters of the Episcopal Diocese in Newark.

Reverend Nathan Wright, Jr. (not to be confused with Willie Wright), is the diocesan head of the Episcopal Church in the city, and he served as chief organizer and chairman of the conference. The Episcopalian parishioners were under the impression that they were engaged in the struggle for civil rights, but Reverend Wright opened the conference with the announcement that "the civil rights movement as we have known it is dead." The net result of the conference, in the words of a reporter for the *Amsterdam News* (the most influential Negro newspaper in New York), was a "confirmation of pride in blackness, in negritude."

Among those who attended the conference were representatives of the "secondary echelons" of the National Association for the Advancement of Colored People (N.A.A.C.P.), the National Urban League, and the Southern Christian Leadership Conference (S.C.L.C.). The top echelons of the Congress of Racial Equality (CORE), the Student Non-Violent Coordinating Committee (S.N.C.C.), and "Paramilitary groups of Ron Karenga's US organization of Los Angeles and Charles 37 X Kenyatta's Mau Mau of New York" were also represented. The *Amsterdam News* also noted, "It was the youthful, energetic, uncompromisingly corrosive posture of the latter type elements that set the tone of the final consensus."

That "final consensus" was expressed in a number of resolutions adopted in secret session. To prevent even a semblance of objective coverage the white press was barred totally, and Negro newspapermen could attend as delegates only, thus binding themselves to accept conference discipline. The resolutions included a call for "paramilitary training," a demand for "rejection of the word 'Negro' for 'black' " and castigation of Christianity as "a white religion that has taken the diamonds and minerals of the world in exchange for the Bible—a bad deal."

It is not yet known whether or not Reverend Wright, an educator as well as a clergyman, supported or opposed this last resolution. But we do know that all resolutions were said to have been adopted without a "dissenting" vote.

The *Amsterdam News* report continues:

> Overriding the attempts at balance of the conservative elements, the conference urged black athletes to boycott the 1968 Olympic Games and professional boxing if the heavyweight title was not restored to Muhammad Ali. Resolutions were also adopted urging partitioning of the United States into "two separate nations, one white and one black"; spurning the military draft and censuring all Congressmen who voted to unseat Representative Adam Clayton Powell, Jr.

Other resolutions called for boycotting publications that accept ads for "skin bleach and hair straighteners"; for setting up black universities to train professional black revolutionaries; and for honoring the late Black Muslim leader Malcom X by declaring a national holiday in his honor. There was also a resolution rejecting planned parenthood and birth control for Negroes as "genocide."

The moderates had some triumphs, however. They won on such issues as "endorsement of black voting, the election of 12 black Congressmen, a nationwide buy-black campaign," and so forth.

The *Amsterdam News* found remarkable the "deep rancor of the delegates towards whites in general and the white press in particular." On Saturday afternoon, July 22, white reporters were forcibly prevented from entering Cathedral House; one white newsman was "physically ejected through a window from a black-only press briefing, and a lone white veteran who briefly picketed the confab's opening day had been belted in the mouth by an unidentified youth who was not arrested."

A picture spread in the *Amsterdam News* of July 29 shows the leading lights of the conference on the podium. They included, in all, William Booth of the New York City Commission on Human Rights; James Farmer, former head of CORE; Chuck

Stone, public-relations director for the conference; H. Rap Brown, then national chairman of S.N.C.C.; Dick Gregory, the comedian and presidential candidate; Ron Karenga, head of Los Angeles' US, a militant Black Power group; Reverend Wright; Omar Ahmed of East River CORE; Reverend Jesse Jackson of the S.C.L.C.; and Ralph Featherstone of S.N.C.C.

After Commissioner Booth's presence became public, Mayor John V. Lindsay of New York was asked if he approved the attendance of a member of his official family at so "corrosive" a gathering of black nationalists. He replied noncommittally that all his appointed officials had enough work to do in New York without seeking extracurricular activities. This comment left the impression that the Mayor had known nothing about Commissioner Booth's attendance at the conference. But the *Amsterdam News* reports that Manhattan Borough President Percy Sutton, a Negro, was "pleased" that Mayor Lindsay had "authorized" Booth's "participation" at the Black Power meeting in Newark. When at a later City Hall press conference a television reporter tried to draw out the mayor on the Booth matter, Mr. Lindsay seemed to "lose his cool." Mr. Booth, however, made no bones about his attendance at Newark: "If this conference were being held in Timbuktu, I would be there." (Booth has since been appointed to a judgeship by the mayor.)

It is worthwhile to recall that earlier in that year, on February 14 to be exact, Rabbi Julius G. Neumann had resigned from New York's Commission on Human Rights and had charged that Mr. Booth was "whipping up animosity among people" and "ignoring discrimination against Jews and investigating only the complaints of Negroes," according to a report in *The New York Times* of February 15. At about the same time Charles Evans Whittaker, retired Associate Justice of the U.S. Supreme Court, charged that Commissioner Booth had advocated a "return to the law of the jungle." Justice Whittaker made this accusation in a lecture at George Washington University after it was called to his attention that Booth was reported to have said, "If the people aren't getting what they need, they should go out and take it."

In one way or another, the resolutions and tactics of the Black Power Conference became the loadstone for much subsequent Negro militancy in the country—and influenced events to a much greater degree than had been anticipated.

Before we depart this riot scene, it is necessary to record that after the black poet and playwright LeRoi Jones was picked up in a police dragnet, a leaflet appeared on the city streets declaring, "It's time all black people in Newark and other states too protest the racist action of the racist Zionist legal system of the country." The "appeal" was issued on behalf of Mr. Jones and his codefendants and was signed by "The Defense Fund," giving P. O. Box 663, Newark, as its address.

Chapter Three

DETROIT, WHERE IT COULD NOT HAPPEN

In retrospect the riots in Newark and elsewhere—even the Black Power Conference—seem but a prelude to the blowup that hit the fifth-largest city in the nation on July 23. Detroit felt the full fury of the mob, and for five days its citizens knew a kind of terror that had not been experienced in this country since the Civil War. "It looked like Berlin in 1945," Mayor James P. Cavanaugh summed up.

But why Detroit? It should be noted that Detroit was one of the few cities in the land where Federal, state, and local officials—business and trade-union groups, social workers, and urbanologists—were convinced that they had found the formula for preventing riots.

A *New York Times* headline reads, "Detroit Was Optimistic." The story goes on: "The boast here was that Detroit was less likely to have a riot than were other cities. 'We've got a lot going for us' was a common statement by Negro and white leaders."

What had engendered this optimism was Detroit's listing in *The Negro Handbook* as one of the "Ten Best Cities for Negroes," and the known moderation of its police commissioner. Detroit's mayor is so liberal that he had evoked wrath in the white community for being too generous with relief handouts. Cavanaugh had also been accused of coddling certain unruly elements and in general of being soft in areas where his critics thought that he should exhibit more firmness.

Detroit hardly ranks as a depressed area. As J. Anthony Lukas of *The New York Times* reports, "Nowhere in the country does a low-income Negro feel so close to affluence, because nowhere is there so much Negro affluence." Describing the speedboat races on the Detroit River, an annual Fourth of July affair, reporter Lukas adds: "This year, at least 30 of the yachts which lined up to watch the race were owned by Negroes. From the shore one could see Negro men in yacht caps sipping cocktails on the fantail while their daughters sunbathed." Also, Detroit has no slums comparable to those in other urban "central cities." Even Twelfth Street on the West Side, where the worst rioting seemed to be centered, is hardly a "slum" as that term is understood by outsiders. It was, not long ago, a prosperous Jewish area, and its "substantial brick buildings, often decorated with ornate marble cornices, are still spacious and solid."

Although Twelfth Street is not what it was in its "kosher days," *The New York Times* reported, never "even in the shabbiest [apartments], does one feel trapped in a prison of asphalt, brick and cast iron as one does in Harlem." Only a couple of blocks away shade trees line the street, and "neatly clipped lawns and hedges" surround "tile-roofed bungalows." And there is no "ghetto," as that word has come to be misapplied to any slum area in the nation. As one white banker put it, "If there's a ghetto here, I live in it."

A Detroit Jewish informant told me:

> I live a few blocks from the old Jewish quarter on Twelfth. It's a good neighborhood. Many fine Negro families live here. I never said a harsh word to them—or they to me. They're nice people. Then when the riots started we were afraid to go out. A Negro neighbor came over to see me. He offered to lend me a gun. I never used a gun in my life and I told him, no thanks. He said he'd lend me the gun to use it when the "black bastards"—those were his words—when the black bastards come around. Go ahead, Barney! he urged. Kill them, they're no good. They don't want jobs. They want loot—your loot. But I said no, I'm gonna move, after fourteen years in the neighborhood. He pleaded with me to stay. But I told him, I'm sorry, I'm moving out. My wife and children are afraid to go out in the daytime now—and that's too much.

Another first-hand account of the riots, this one written by Saul S. Friedman and published in *The National Review*, tells of "the burning out of the lone pharmacy on Hough Avenue between East 55th and 105th, *merely because it was operated by Jewish brothers* (who, incidentally, had been in the area for more than twenty years)." The emphasis is mine.

Detroit Negroes find work in large numbers in the integrated automobile industry as members of the United Auto Workers Union. They earn an average of $4.50 an hour; are highly involved in local community activities; have elected two Negro congressmen, achieving the highest Negro representation from a single community in the whole nation; and have integrated leadership to help prevent riots. Then the riots came, and all the hopes and "carefully laid plans collapsed," as a reporter for the *Times* put it.

Why? From *The New York Times* of July 25, 1967, we learn that "At police headquarters yesterday afternoon, Robert Tindal, the executive secretary of the local National Association for Colored People [sic], warned that unless force was used soon to break up the mob there would be much more serious trouble. He was proved right."

The Reverend Nicholas Hood, Detroit's only Negro city councilman, said that he was moving out and remarked about the riots, "It's not a matter of civil rights. It's just lawlessness—out and out lawlessness."

"We don't want no reverends talking for us," one Negro shouted.

"I drove up Twelfth early Sunday morning," said Richard Marks of the city's community-relations bureau. "The looting was going on, and I didn't see the police."

A Negro boy watching the looting reassured his companions that nothing would happen. "They won't shoot," he said. "Mayor Cavanaugh said they aren't supposed to shoot."

Why?

Mayor Cavanaugh said they aren't supposed to shoot. "They" are the police. But the Mayor's permissive approach to the wanton behavior of the mobs in the streets scandalized many law-abiding citizens of both races in the city.

Bitterness between Cavanaugh and his police department had been permitted to fester. It should be recalled that when Cavanaugh took office he dedicated himself to helping to resolve certain problems in the troubled city. When his predecessor, Mayor Louis C. Miriani, had ordered a crackdown on terror after a number of white women had been raped in Negro neighborhoods, Cavanaugh had aroused the massive Negro support necessary to elect him by promising to ease tensions in areas of racial bitterness.

After his victory the Mayor appointed a well-meaning but ineffective professor as police commissioner, but he was soon replaced by the then incumbent, Ray Girardin, who was to carry out, with perhaps a bit more firmness, his predecessor's policies. I am informed *his* heart was never really in his work either. Soon the mayor had incurred the wrath of almost the entire law-abiding citizenry. And, as if that were not enough, "he mumbled and he fumbled," as one Detroiter put it, and he was thus thoroughly unprepared for the tempest when it came, although the first signs could have been detected without special meteorological skills.

But almost all the power establishment, as the Negroes like to call it, was convinced that *nothing* worrisome could happen in a city where there was little unemployment; high wages; good neighborhoods, by and large; and lots of money for antipoverty programs. James P. Cavanaugh believed in all his nostrums, and all his sociometric indicators told him (pragmatically) that all was well because everything seemed to function well.

In the days after Federal troops were called out Detroit was to experience the full fury of the mob. This matter of Federal troops later caused a political hassle between Governor George Romney of Michigan and President Lyndon B. Johnson. Was Johnson remiss in not answering the Governor's demand for troops posthaste? Or had the Governor vacillated over making the request while Detroit and half a dozen other Michigan cities burned? Certainly the destruction of vast areas of the fifth-largest city in the land was not going to keep two of our leading politicians from playing politics as usual. So Detroit burned. And before the fires were put out more than two score people were dead, 1,000 had been wounded, and more than 3,000 had been arrested. And between a half-billion and a billion dollars' worth of property had been looted or destroyed.

The looting was not only in retail shops—a pair of shoes here and a television set there. Some of it was on a wholesale scale, as trucks backed up to remove pool tables and jukeboxes too bulky for ordinary car removal. In the riot areas everything not marked "Soul Brother" or "Friend of Soul Brother (with "Friend of" etched in small letters beside "Soul Brother" scrawled large) was either burned, stolen, or smashed. Almost none of the Negro or "soul" businesses in the former Jewish section of Twelfth Street was destroyed.

Was there outside influence in Detroit? This question seems to annoy the officialdom of our urban centers as almost no other question does. Especially when the question is phrased to read "outside agitators," denial becomes almost ritualistic. This reaction is sometimes explained on two grounds: that no outside agitation is needed to spark "rebellion" and that local authorities are competent to deal with "purely local" matters, as they are called. In any case, as post-mortems show in Detroit and as will be found elsewhere when proper studies are made, outside agitation did indeed help to "fan and guide the violence." Two masked (for protection) Negroes made this point clear over NBC television on the evening of September 15, 1967. One of these men, identified only as Mr. X, his face concealed by a stocking slipped over his head, told of cellar meetings of militants at which they were indocrinated by two agitators, a young man and a woman, who told them "to break windows, get other people to do likewise, to loot and to burn buildings."

He told how he was sent on another night to an East Side house, where a white man handed him a rifle after he had identified himself as one of those who had attended the meeting and who had participated in the rioting. Later in the program the second anonymous Negro, who was identified as a member of the "inner circles of black extremists," claimed that the next time they would attack the police station. Asked whether or not the top leadership of the group was composed of black or white men, he replied, "It's the same color you are brother." In this instance the "brother" spoken to was a white television reporter.

The Henry brothers—Milton, and Richard—had long been active in fermenting a kind of abrasive "blackism" that surely left its imprint on the least stable elements in the Negro community. Milton is a lawyer, and Richard works for the government. They are articulate and totally committed to a separate Negro nation in the South for black men only. They now head up the Republic of New Africa (RNA).

A leading Negro Marxist theoretician also lives in Detroit. James Boggs regards black rioting in the streets as fulfillment of the revolutionary vision of armed men tearing up the fabric of our civilization. He sees the city as the Negro's property to take over and to control. He had hailed the Watts riots as a "major battle" testing the wills of "18,000 [white] soldiers and the black people." He has helped to make Detroit, despite all its welfare programs—and they have been prodigious—the black-nationalist capital of the United States.

Meanwhile, special bulletins kept splashing on the television screen, telling of rioting in Rochester, New York; Cincinnati, Ohio; Portland, Oregon; Pontiac, Michigan; Englewood, New Jersey; and Cambridge, Maryland. The last one could have been predicted even by Mayors Cavanaugh and Lindsay. It was then July 23.

About a month earlier Mrs. Gloria Richardson Dandridge, a Negro militant known for her ability to whip up the enthusiasm of local crowds, arrived in Cambridge. Some worried citizens thought that it would be a good idea if she were asked politely to depart. But the city officials, including the chief of police, said no. So they awaited the arrival of Student Non-Violent Coordinating Committee leader H. Rap Brown, fresh from his triumphs at the Black Power Conference in Newark.

Once arrived in the city, on July 24, Mr. Brown made a speech. In it he confided that "No one has to tell me who my enemy is and I know how to kill him. . . . No one has to tell me how to kill my enemy. . . . When I get mad, I'm going out and look for a honky [white man], and I'm going to take out 400 years' worth of dues on him. . . . In a town this size three men can burn it down.That's what they call guerrilla warfare. . . . You should have burned this town down." After this exhortation from the leader of a "nonviolent" movement, the mob went about burning the town down. Mr. Brown was wounded slightly, treated in the local hospital (which had not been burned down), released, and arrested on a charge of inciting to riot.

On July 25 the national guard was called out to quell the riot in Cambridge, and Mr. Brown departed from Maryland in haste, proceeding to the nation's capital, where he was arrested on a Maryland warrant and released on $10,000 bail. Then he held a press conference, at which all the nation's press eagerly awaited to report his every word. Television cameras assembled to carry his message to every home in the land, a message urging Negro militants to "kill the honky," by which Mr. Brown meant "kill the white man." Asked whom he had in mind, he replied that maybe he meant "Lady Bird," or the questioner himself.

He surfaced again on August 6 in New York City's Borough of Queens at a mass rally on behalf of seventeen Negroes charged with conspiracy to murder several moderate Negro civil-rights leaders. He said at that time that the recent riots had been only a "dress rehearsal for revolution." The white press, as had become unprotested custom, was ejected after having been "invited," but the reporters took it good-naturedly. To ensure that the alleged conspirators received full justice, Mr. Brown urged his followers to appear in court "with a show of force. That's black power," he added. The rally also heard some unflattering remarks about Uncle Sam and Uncle Shylock.

In Cambridge the frenzied mob had looted and burned with such passion that even "Soul Brother" Hansell Greene's motel was destroyed: "I'm broke. I'm beat, and my own people did it." Perhaps Greene's misfortune need not have occurred if Police Chief Bruce Kinnamon had been quicker to deploy his men and had not tried to temporize with a mob bent on violence. Who knows? Less than a month after the Cambridge riots Mr. Greene killed himself, the first genuine, as distinct from the psychological, suicide. Negro psychologist Kenneth Clark has attributed the riots to a desire for self-destruction. No, Dr. Clark. The only real suicide was Greene's. The hoodlums and innocents who were killed *because* of the riots had no desire for *self-destruction*. Mobs never do—neither here nor anywhere else where men go into the streets to redress grievances, especially where due process is available.

Chapter Four

SOME FUN IN FUN CITY

On July 23, 1967 John V. Lindsay, the unflappable Mayor of New York City, arrived early and smiling from a summer vacation in Miami in order to help calm a small riot that had broken out the previous evening in East Harlem. A few rocks had been thrown, a car or two overturned, a few store windows shattered, a couple of fires set, and a few bystanders injured. When he heard of it, the Mayor declared, "It's not a riot; it's only a demonstration."

On the following night more trouble broke out in the teeming Puerto Rican section of the city known to local residents as "El Barrio," and the police were called out in force, but with instructions not to act tough. After these instructions and after a talk between the mayor and some neighborhood leaders, the rioting took on extra dimensions. Smashing of store windows and looting were widespread; two people were dead and many wounded by the time a driving rainstorm in the early morning hours of July 25 brought a halt to the "demonstration."

Suddenly the Mayor was less available for comment. On television he looked grim, and for the first time since he had taken office a faint shadow of doubt seemed to cross his brow. Maybe giving in to the bullying toughs and playing Mr. Clean to a dirty city were not the way to run the ethnically divided metropolis of which he was chief executive officer. In all fairness to Mr. Lindsay, nothing in his background had prepared him for a Stokely Carmichael or an H. Rap Brown. Someone once remarked of the late Justice Oliver Wendell Holmes that nothing in his past had prepared him for Hitler.

A couple of days later New York saw the omens of future rioting. No more (it seemed) would black children of the "ghetto" riot and destroy their own neighborhoods. There were bigger and better stores to vandalize downtown, on Fifth Avenue, for example. About 100 of them came to the lushest shopping street in Fun City, straight from Central Park, where they had been enjoying one of the Mayor's summer festivities, and smashed windows and looted some of the poshest shops on the Avenue in full sight of home-going theatergoers. A similar incident took place in the fall. And in December, 1969, Adam Clayton Powell would point to "downtown" as the place for the riots next time.

The "demonstration" that the mayor had played down the previous Sunday seemed to promise bigger things to come. It seemed to stir the Mayor, who had worked so hard to pour money into all kinds of projects to assure New York City a peaceful summer. He began to walk coatless through Harlem, hugging and being hugged by hoodlums and roughnecks in the area. The Mayor pledged that he would fight the blight and poverty that he was sure were causing all the mess, apparently if it meant spending the last cent in the city's, the state's, and the Federal government's treasuries. Of course, the television cameras always accompanied him on his trips.

One day he paid the Staten Island ferry to take his "charges" up the Hudson River; the next day he publicly humiliated the 95 percent of Harlem's law-abiding residents by giving an airplane ride to the area's "five percenters." (A term coined by social workers who regard them as the 5 percent whom no amount of reason, only bribes and dope, can influence. Black militants say it refers to the 5 percent who will make the revolution.)

In East Harlem, a local politician, Councilman Low, went with a television crew to visit one Mrs. Santiago's cockroach-infested home. The politician and camera crew departed in haste when she responded to the question of who was responsible for the rioting: "The bums on relief made the riots. They got plenty of time for them." Some say she mistook the Councilman for the exterminator, others that she was ungrateful for what television was doing for her. But Mrs. Santiago only wanted someone to get rid of the cockroaches for her. The television crew and councilman had other things on their minds.

Not far away Mr. Harvey Helpman found himself with only $500 worth of merchandise the day after the nonlooting "demonstration," despite the Mayor's statement that there had been almost no looting. The day before the "demonstration" Mr. Helpman had inventoried his stock at a value of more than $10,000. Now he decide to call it quits. "I'm leaving tonight [July 26] and not coming back," he said.

The charge of "compromising" with the criminal element in "Fun City" is not only—as is usually thought—a "white backlash" complaint. The responsible Negro community is even more concerned with the problem, as the following editorial reprinted in full from the September 30, 1967, issue of the *Amsterdam News* makes crystal clear:

> Someone may have optimistically called New York "fun city" but there are too many working overtime to kill the fun.
>
> Lovers and a diplomat are brutally beaten in Central Park. . . . Policemen are permitted to moonlight as cabdrivers because hackies have become holdup pigeons. . . . Pedestrians are mugged. . . . Bus riders are alarmed as bandits invade buses, threaten two drivers and wound one of them.
>
> Whether it is summer heat or chilly fall—no matter the season—these lawless vultures plot crime faster than our sociologists and urban affairs specialists can compile statistics.
>
> Before the Central City is condemned as rotten to the core, New Yorkers must demand the eradication of crime just as vigorously as we clamored for rat control.
>
> We must call for restoring the ligitimate, unbiased use of firearms by our police, the return of the right of a man to defend his home against robbers and demand more concerted effort against criminals, white or black, old or young.
>
> We can't get rid of crime by ignoring or compromising with it.
>
> And we can't use slingshots or statistics to fight animals bent on killing. We don't have time.

Leading political spokesmen, including former Governor Romney of Michigan, Joseph Clark of Pennsylvania, Senator Charles Percy of Illinois, the late Senator Robert Kennedy of New York, and Mayor Lindsay, have expressed the fear that urban rebellion menaces every metropolis in the land. Black militancy seems to increase with every appointment of a prominent Negro leader to public office. This paradox cannot be explained away by psychological jargon, nor will wishful repetition of the refrain that until poverty is completely eliminated the poor (that is, Negroes) will continue to be frustrated.

Attacks on lives and property in the so-called "ghettos" of the United States can be prevented only by a society reoriented to respect and defend the private citizen in his pursuit of peace, rather than the criminal in pursuit of loot or rebellion. Yet the courts in recent years, and I think that every district attorney in the nation recognizes this change, have set themselves up not as arbiters of justice but as social instruments to weigh the rights claimed and the wrongs said to have been suffered by particular groups and individuals in much the way of people's courts in totalitarian lands.

Proper and effective police power—the kind that the *Amsterdam News* calls for—views every act of vandalism or lawlessness without regard to color, which is how the police must approach everybody, even in our color-conscious society; it is one of the surest ways to solve the problem of urban violence. Massive doses of permissiveness and money will not do it!

Chapter Five

THE PRESIDENT APPOINTS A COMMISSION

By August 1, 1967, more than 100 American cities had endured rioting mobs in one form or another, including most of the "ten best cities" for Negroes—Hartford, Sacramento, Minneapolis, Houston, Columbus, Portland, Oregon, and Detroit. President Lyndon B. Johnson therefore announced the appointment of a commission to study and report to the nation the causes and cures of the riots of 1967 by no later than midsummer 1968. The President asked his newly appointed commission to answer three questions: What happened? Why did it happen? What can be done to prevent it from happening again?

Answers to these questions did not require a year of study by a fifteen-man commission. The simple fact is that, if the President did not know *what* happened, no amount of research can tell him or the nation the answer to this preposterous question. It has been charged that the reason it was asked was entirely political. The president did not want to call the riots by their rightful name. He did not want to use the word "insurrection." Nor did he care to characterize the events of the summer of 1967 as "unruly" or harmful to the safety of the nation because that might alienate some voters during the Presidential election year 1968. It *could* also be that he simply did not know. Nothing is impossible it has been said.

In any case, those questions were posed, and these men and a single woman were appointed to seek out the facts and to make recommendations: Mrs. Katherine Graham Peden, Kentucky Commissioner of Commerce; Roy Wilkins, Executive Director of the National Association for the Advancement of Colored People, a Negro; Republican Senator Edward W. Brooke of Massachusetts, also a Negro; Cyrus R. Vance, a Presidential aide; Attorney General Ramsey Clark; Governor Otto Kerner of Illinois, named chairman of the commission; Mayor John V. Lindsay of

New York, named vice-chairman; Charles B. Thornton of Litton Industries; Representative James Corman of California, a Democrat; Vice-President Hubert H. Humphrey; Herbert Jenkins, Atlanta police chief; I. W. Able, President of the United Steel Workers of America; Representative William McCulloch of Ohio, a Republican; and Senator Fred R. Harris of Oklahoma, a Democrat.

The President subsequently announced the appointment of David Ginsberg, a Washington lawyer, to serve as Executive Director of the Commission. This appointment came about after someone called the attention of the Chief Executive to the fact that the minority that had suffered the most property damage in the riots had not been represented at all.

Let us note, too, that on the commission there was not a single Negro who can be said to have "made it" in terms of the American dream. Roy Wilkins, who is a professional dedicated to the welfare of his people, is a special case. Senator Brooke had to serve in the humiliating position of the other "Negro" representative, something he has never before done during all his public life. He was not elected as the *Negro* Senator from Massachusetts but as the Republican Senator of all the Commonwealth. There was actually only one Negro—as spokesman for Negroes—on the commission.

Why did the President not choose a Negro business man? Or a professional man who would reflect the dignity of Negro accomplishment? After all, there *are* such distinguished and respected men in the Negro community, although not many of the former. That was all the more reason to have such a representative among the politicians with whom the committee was weighted.

Finally, there was not one man on the commission who had ever given the subject any serious thought—in terms of direct knowledge or immediate confrontation or special understanding. But the President is to be congratulated for avoiding the appointment of some jargon-spouting academic. For the academy has done more to exacerbate relations in this volatile area than any other institution in the nation. J. Raymond Jones, a Negro and former head of the New York County Democratic Committee (Tammany Hall) once said, "Harvard has done more harm to the Negro than bad booze."

After the appointment of Ginsberg and a Negro technical staff member, the commission was ready to investigate the following specific questions, as outlined by the President's staff:

Why do riots occur in some cities and not in others?

Why does one man break the law, while another, living in the same circumstances, does not?

To what extent, if any, has there been planning and organization in any of the riots?

Why have some riots been contained before they got out of hand and others not?

How well equipped and trained are the local and state police, and the state guard units, to handle riots?

How do police-community relationships affect the likelihood of a riot, or the ability to keep one from spreading once it was started?

Who took part in the riots? What about their age, their level of education, their job history, their origins and their roots in the community?

Who suffered most at the hands of the rioters?

What can be done to help innocent people and vital institutions escape injury?

How can groups of lawful citizens be encouraged, groups that can help cool the situation?

What is the relative impact of the depressed conditions in the ghetto—joblessness, family instability, poor education, lack or motivation, poor health care—in stimulating people to riot?

What federal, state and local programs have been most helpful in relieving these depressed conditions?

What is the proper public role in helping cities repair the damage that has been done?

What effect do the mass media have on the riots?

The kind of report that the commission was expected to produce was clear. Most of these questions are merely rhetorical, containing hints at the desired answers. The report would contain what most of the commissioners already believed. Mayor Lindsay was sure, even before he had had an opportunity to spend one full day with the commission, that the riots were all caused by "poverty" and "conditions in the ghetto." A *New York Times* editorial of July 30, 1967, which concluded with the hope that the Commission's work would in time produce "a product that will be a better and a color-blind America," suggested the preventives that the commission would recommend.

When the report of the National Advisory Commission On Civil Disorders (the Kerner Report) came in, it was hailed by all good white liberals. Bayard Rustin was not so enthusiastic, and Kenneth Clark who said he'd seen dozens of reports in his lifetime, held out little hope that this study by a Presidential commission would achieve more than the others had. Rustin said: "It helped no one by blaming white racism for the riots."

In concluding that white America was racist, the Commission—established to seek out the causes of the 1967 riots—thereby laid the responsibility for the riots upon the consciences of white America. Editorial writers and the mass media's pundits whipped themselves into masochistic frenzies following the Report's release. But the people, on the whole, refused to swallow it, for its bleak message that America was fast becoming two nations—one black and one white and both unequal—was at once too simple and too complex. Certainly it did little to prevent the orgy of violence that swept the nation following the assassination of Martin Luther King, Jr. And, probably it helped bring on George C. Wallace's backlash candidacy in the 1968 election campaign.

One weakness of the Report that remains unexplained is its failure to interview any of the hundreds of storekeepers whose shops were either looted or burned. The lootee—the owner of the "joosh" stores as LeRoi Jones put it—was the forgotten man in the aftermath of riot.

Chapter Six
"COLOR-BLIND"

Let us pause a bit to examine the concept of "color-blindness." A half-century ago Americans still believed, innocently, that almost any problem could be solved by passage of the correct law or adoption of the correct attitude. All decent people held similar beliefs, especially in the area of race relations, and most particularly in the North. Men of good will accepted the liberal theory that the sooner questions about race were obliterated from public life—from job questionnaires, civil-service tests, newspaper accounts, even from the Census—the sooner all Americans would become "one people indivisible, with liberty and justice for all."

It was a time that could elicit such a brave statement as that of the distinguished anthropologist Franz Boas, who said that it would be a sad day for the Negro if he were ever to get, or go after, a thing just because he was a Negro. That, Boas felt, would not only be demeaning to the Negro's individual dignity; it would also be unworthy of his race.

Total integration of Negroes into white society was the dream of almost all men devoted to solving this troublesome problem. And, in the North at least, most men of good will thought that they had achieved such a balance. Negroes could go to the same movie houses almost everywhere, to the same neighborhood schools, to most restaurants, and to many a good hotel. Their efforts were all in the right direction, and although things did not move quite fast enough to satisfy many in the Negro population, for the next fifty years no one seemed inclined to rock the national boat. In 1954, when the Supreme Court ruled against school segregation in *Brown* v. *Board of Education*, Negro psychologist Kenneth Clark hailed the decision as the triumph of "color-blindness" in the United States.

Then disillusionment set in. Suddenly the white man became conscious that he had never really looked a Negro in the eye as a friend, never spoken to a Negro man to man, never attended a party at which a Negro was a guest, never invited a Negro to his own home. This discovery was especially widespread in the North. The northern white had lived in an unseeing relationship with the Negro in his midst. Not that the New Yorker, for example, did not often go to Harlem for a good show, "real" southern-fried chicken, or "to change his luck," as he sometimes put it with a leer. But essentially he knew the Negro as the elevator boy, the postman, the singer, the musician, the comedian, the some-*thing* or other. Rarely did he know a Negro *personally*.

Things had been going along relatively smoothly, or so the northern white thought, and in large measure he was correct. Then, the highest court in the land made Negroes "visible" to millions of people who had never given them much thought one way or the other.

Suddenly, newspapers that had almost never used the word "Negro" in years, became so self-conscious on the subject that they simply did not know where to stop. Negro civil-rights groups like the Congress of Racial Equality, which had been around for some years before the Court decision and which hardly anyone had ever heard of, suddenly came to the fore. New organizations, like the Southern Christian Leadership Conference headed by an unknown minister, Reverend Martin Luther King, Jr., appeared on the scene.

People started marching and singing, and one or two old-line Negro labor leaders—actually there was only one, A. Philip Randolph, of the Pullman Porters Union, who was competently assisted by Bayard Rustin of the A. Philip Randolph Institute—intensified campaigning for civil rights. Even the term "civil rights" had a new ring. Soon a huge march on Washington was organized by Mr. Rustin. Tens of thousands of people, white and black, marched in orderly procession, holding hands and singing "We Shall Overcome." And everybody was happy, and everybody was gay, and the editorials rejoiced in the dawn of this new day of freedom.

Then the whole thing blew up in the faces of the men of good will, as younger Negroes, somewhat better educated than the preceding generation and better subsidized by the government and their parents, revolted against the condescension and paternalism they detected in some white attitudes toward Negroes in general and toward some Negro leaders in particular. They knew that when one of their older spokesmen made an ass of himself on television and said things so preposterous as to make one wince, the white reporter, who would have laughed out loud had he heard it from a white man, would sit there saying, "Yes, Mr. This, and Really, Mr. That" and so forth. It sounded very much the way a southern white gentleman would talk down to his colored neighbor. But at least in the South the Negro *was* his neighbor. Probably because of this outrage at the patronizing attitudes, especially of Jewish liberals who were always in the forefront of any civil-rights action, even ahead of the Negroes themselves, Negroes started to go it more and more alone. One Negro youth organization, the Student Non-Violent Coordinating Committee—S.N.C.C. or "Snick," as it is known—which began as an adjunct of the nonviolent movement of Dr. King, became more and more insistent on a black identity. Its members no longer sought to persuade the nation to become color-blind, or to act as if it were. Soon the Congress of Racial Equality (CORE) adopted the same stance, as did a number of the black nationalist organizations that had sprung up across the land. "Look at us. We are black and beautiful" they said. Stokely Carmichael and others proclaimed the new semantics of race: "We are *black*, not Negro or colored." Disdain for the white man was publicly expressed. Terms like "Mr. Charley" and "Honky" for the white man in general and "Goldberg" for Jews in particular enriched their rhetoric. Soon these young blacks were screaming for a kind of separatism that had not been seriously considered since the mid-1930s, when American Communists tried to curry favor with black nationalist elements by calling for a separate Negro republic in the "Black Belt." Now some Negroes wanted to go it alone, and they wanted the white world to be made aware of their own new self-awareness. Yet the liberal kept trotting out the old tarnished "color-blind" image as his goal for the future. The militant Negro said no. And his disturbing voice was more and more frequently heard in the land, regardless of what "moderate" leaders might say or do.

What are the merits—if any—of a totally integrated society which, properly understood, is a color-blind society? Possible or not—and it has certainly not proved possible in this country—is it desirable? The answer which seems to suggest itself is that it is not. Not in the United States, certainly, with its mixed population made up of ethnic groups from the whole world.

Certainly Nathan Glazer and Daniel P. Moynihan proved in *Beyond the Melting Pot*, to the extent that they proved anything, that minority groups tend to stay together in homogeneous *neighborhoods* for longer periods of time than anyone had previously thought likely or desirable. The dream had always been the *melting pot*, a term taken from a play by Israel Zangwill. In any case, the pot was not hot enough to melt together Italians, Jews, Swedes, Chinese, Japanese, and various other ethnic or national minority groups. Now this news was hardly a revelation. Anyone with eyes had known it all along. What made the Glazer-Moynihan study front-page news was their belief that it was a pretty good thing that American society had not been completely homogenized. Suddenly terms like "pluralistic society" and "open society" took on new meanings—although a few professors, members of certain liberal arts faculties at eastern universities, suffered some anguish when the Glazer-Moynihan findings were announced.

Soon the word "homogeneous" was associated with the pejorative "homogenized," and the term "color-blind" was arousing strong negative reactions among black men. From the Negro's point of view integration *is* a white man's problem. The black man tried desperately to join the white man in a common pursuit of happiness. He did try to integrate himself into white society. Perhaps he was mistaken in the attempt. But *he* had not raised the question of integration; he had not wanted to be color-blind; he was not especially interested in marrying the white man's sister.

All the arguments and solutions for the American Negro problem current today are white-inspired and liberal-inspired. That Negroes (with the exception of the great Booker T. Washington, who knew that the white man's fine talk was a snare to be avoided as if it were a bear trap) lacked the leadership to counter such pious white attitudes and had failed until the mid-twentieth century to build for themselves institutions of self-help and, yes, self-indulgence, is regrettable. But decent white liberals insisted on promising them something that they could not deliver—and probably never will be able to deliver—and now the Negroes know it.

I do not argue that had Negroes built their own institutions in their own neighborhoods and communities they would be economically better off today than they are. After all, not all minority groups in America have had the same rate of achievement as has every other group. Some climb the socioeconomic ladder more rapidly than do others. Some—less "achievement-oriented"—do not. So what? It is not necessary for a society to be so democratic and fair that each citizen will be equal in every respect to every other citizen. This goal is unrealizable; at best utopian, at worst totalitarian. It is enough that each of us is equal before the law, that and no more.

The law can no more make a man love his neighbor than thousands of years of the Judaeo-Christian ethic could. It has been pointed out that the ideal of total integration is the mirror image of South African *apartheid*. Integration demands that all blacks be brought, even by force, into the mainstream of white society; *apartheid* requires that all blacks be forced out of the mainstream of white society. Both are indecent because both are totalitarian in concept. Yet the South African measures have for their stated purpose, not so much the maintenance of a kind of racial purity, but self-defense, because blacks outnumber whites there by huge proportions. The same is not true in the United States.

Yet even in this country we are beginning to hear of separation as a desirable goal, and it is mainly being preached not by racist white but by blacks. To be color-blind in the United States in the 1960s is to be blind to the realities confronting the nation. And Dr. Clark, who had hailed the Court decision of 1954 for its "color-blindness," was ready only a half-dozen years later to say, "I must confess bluntly that I now see white American Liberalism primarily in terms of the adjective 'white.' "

Chapter Seven
BACK TO THE ACADEMY—AND THE RIOTS

It is worth returning at this point to the opening paragraph of Chapter One, in which I described how social scientists at the Lemberg Center for the Study of Violence had predicted that certain cities—Boston among them—would be unlikely to suffer violence in 1967. These social scientists thought they had learned how to avoid riots in the "future." One might have imagined that after such a sociological gaffe they would fall over dead, or at least apologize to whoever subsidizes their banalities. But no. A few months later, after the riots were "over," they were at it again, this time telling us that they knew just why the riots occurred.

In the opinion of John Spiegel of the Lemberg Center, the riots were caused by the massive "in-migration"—note the jargon—of Negroes from the South to the North. Unable to cope with northern folkways and culture, they just naturally rioted, or words to that effect. Of course, the press was duly impressed, as it had been earlier in the year when it reported the Center's first findings on the subject. For many academicians, however, it is the pious predictions of the day, not yesterday's errors, that count. Thinking to humanize their mistakes, they repeat to themselves, "To err is human, to err is human."

Dr. Robert Coles of Harvard University attempted to illustrate the newest Center findings; he quoted extensively (and anonymously) from a "series of one"—that is, a family from Alabama whose members, although they had not participated in the latest Boston rioting were prepared to do so the next time around. He permitted the mother of eight children to ramble on; and she told the tale that Dr. Coles wanted to hear. He also included the remarks of an uneducated son of twenty-two who spoke the sophisticated lingo of hate and aggression borrowed by Stokely Carmichael from the incendiary works of Frantz Fanon.

In any case, some of the things Dr. Cole's informant told him are of interest; for example, "You don't have friends up North, you just see a lot of people." We learn how her husband has to "hide out" so that she can receive relief. That her husband is almost compelled to disappear so that local welfare and federally supported A.D.C. (Aid to Dependent Children) funds will be provided for her family's maintenance is taken for granted. Even if her husband had a viable skill and could earn the prevailing skilled laborer's wage in Boston, it is doubtful that her family would see as much "cash money" each month as she receives from "the welfare." (The informant told Dr. Coles that she never in the South saw in any one year as much cash as she received each month in relief payments in Boston.) But Dr. Coles is apparently not interested in that aspect of the subject, only in the fact that his family fulfilled the requirements of the "in-migrant" thesis. Yet any careful study of the cities and suburbs where the rioting was hottest during the year 1967 *fails absolutely* to bear out this thesis.

The facts are—and they were available to Dr. Coles and to *The New York Times*, which printed his "findings" on September 17, 1967—that a survey conducted by the Detroit *Free Press* and sponsored by the National Urban League showed that, although only 23 percent of the nonrioting Negroes were native Detroiters, *46 percent of those who participated in the riots were born in Detroit.* As the *Free Press* summed up, "This explodes whatever remained of the theory that race riots are caused by Southern Negroes who can't adjust to the pressures of big-city life."

Apparently the explosion was not heard inside the walls of Harvard or Brandeis.

At Columbia University, which is situated on the fringes of the Negro community in Harlem, a hassle over erecting a gymnasium on the campus brought instant retaliation from angry black men who thought that they should be consulted on the matter since they now regarded Columbia as being part of the community. This confrontation was abetted by protests from the Negro students registered at the University. Later, white and middle-class students of the radical left—the Students for a Democratic Society (S.D.S.)—joined the fray, with demands for student control and student power, but they were not joined in this by the young Negroes, except that the latter demanded their own separate quarters on the campus; and that the University start emphasizing African and Afro-American studies in the curriculum.

This had become almost a common demand of black students in almost all schools of higher learning. What brought the Columbia fracas to national attention was not the original protest of black students, but the much more abrasive behavior of the leaders of the S.D.S. who forcibly occupied various faculty buildings, the dean's office, and student halls. They set about creating such havoc that eventually the police were brought in to forcibly evict them.

What was interesting to observe was that the building occupied by the black students was almost totally devoid of vandalism, while the white—and infinitely richer—students of S.D.S. behaved the way most "experts" would have predicted from only the less "advantaged" elements of the student body. What is significant here is that it proved once again that lack of money—poverty—is not a gravamen for the behavior of the mob, any mob, rioters in the streets or well-to-do students. Rooms were defiled, books mutilated, furniture broken. The mystique of rectitude is in itself a destructive force and is not inhibited too much by family background or papa's bank account. The Negro student at Columbia is there for an education, and if he thinks that part of student life is private group life, he has a point there. Fraternal organizations, clubs, special religious and ethnic group activities are familiar adjuncts of university life. That some young blacks think that the way to achieve greater group identification is by way of a total segregation of social (and other) activities is understandable. This notion shocks only the liberal who thinks, ritually, that Negroes should know better than to do as others have done before him. So long as the Negro student wants to remain separated from his white classmate for any reason whatsoever, he certainly has every right to do so. The right to free association must not be denied to anyone; so long as he does not violate the privacy of others, his privacy must indeed be protected. That this will not help toward the integration of the races—especially on the higher scholarly levels—is unfortunate. But the new black student must be allowed to experiment with his new-found freedoms, including the one that allows him to behave as if he had no freedom at all. So long as it does not impinge on the liberties of others, let him even play at being African if he wants to.

(However, when the university fails to act responsibly as at Cornell where black students brandished guns without being summarily disciplined, that is a sign of the failure of the university—not the black students. They took the bad habits of the slums with them—and the college fathers told them, in effect, that the university was, indeed, an extension of the slum, not a place of higher learning. The lesson for Columbia, and for other institutions situated close to the "ghetto," is clear, and presents a danger to the fragile relations which exist between students and faculty in even our best universities.)

I mention the Columbia experience because even reputable academics practicing the nebulous disciplines of the social sciences, are frequently responsible for the transgressions committed by later generations of students, who have interpreted the professors' teachings to suit their own time and temper. Thus, when Dr. Daniel Bell, one of the most moderate and level-headed professors at Columbia, tried to tone down the meaning of the present turbulence in our society by pointing to past outrages—always finding them greater than today's—he was perhaps seeding the kinds of student *schrecklichkeit* Columbia University and other institutions of higher education witnessed in 1968 and 1969. Dr. Bell seems to believe in a kind of sociological one-upmanship. If one complains of noise in the streets, he will trot out statistics to show that it was noisier in 1942 (or was it 1492?). Are our streets unsafe to walk in? They were a lot less safe in 1863, says Dr. Bell. In any case, he seems set on proving that what was once was worse. In this he personifies the absolute modernist; nothing in the past holds a candle to the shining new present, and the newer a thing is the better. Dr. Bell understands that it is not poverty *per se* that is responsible for social upheavals, but rather the kind of fatalism it induces. "*Social tensions*," he says, "are an expression of unfulfilled expectations." This fact has been known to every ordinary trade unionist for a long time and to every radical before and after Nikolai Lenin. Dr. Bell explains further. "It is only when expectations are aroused that radicalism can take hold." Of course. But the question remains, why should unrealizable expectations be instilled in the minds and hearts of people?

After all, not all desirable goals can be achieved by any man. Even if socialism is a desirable goal, it is still not certain that it will be fulfilled in democratic terms, regardless of the claims of certain special pleaders. It is well to note that "democratic" socialism is usually promoted by men born into Jewish families, men like Dr. Bell, Sidney Hook, Irving Howe. No other names come to mind. When spoken by a Stokeley Carmichael, his black "socialism" may be a time bomb with the mechanism broken, so that no one, least of all Mr. Carmichael, knows exactly when it will go off. But, when the nation faces a predicament, Professor Bell tends to identify it with his own, with the alienation experienced by some American intellectuals. He may not perhaps realize that, almost without exception, the names he cites to prove his point are all Jewish names—Harvey Swados, James Wechsler, Saul Bellow, Leslie Fiedler, Alfred Kazin, and Richard Hofstadter. These men, with the possible exception of Dr. Hofstadter, may indeed all be alienated both from American society and from their Jewishness. It would seem Dr. Bell is simply trying to reassure his middle-class, liberal, and well-insulated friends that our current troubles will turn out all right. But Dr. Bell may be wrong. It may turn out that the American Negro, or in current terms the black man or Afro-American, will not be satisfied to burn his "ghettos" down but will go after campuses located near the slums.

I have devoted this much attention to Daniel Bell because he thinks that violence in the streets probably represents a form of "unorganized class struggle." That is what Mr. Carmichael and H. Rap Brown learned from his last tome, even if they did not read it. What wisdom will they learn next? The defense of criminality as one form of *embourgeoisement*—what a word!—of certain immigrant groups as they ascend the ladder of opportunity in the United States needs no refutation. The pity is that Daniel Bell, the compleat pragmatist, has never learned that "the criminal deserves his punishment" because he *needs* it. In defense of crime "as a form of resentment," Bell implies that it is some kind of variant of the "class struggle." And class struggles, regardless of the forms they take, are the proper study of modern sociologists. H. Rap Brown's announcement to a white crowd in Queens that the 1967 riots were only "dress rehearsals" for the real thing sends shivers down one's back, for it is possible that Mr. Brown may yet find sanction for his hatred in the innocent remarks of a sociologist at Columbia University. (Dr. Bell has since departed for Harvard.)

In mid-summer 1968, New York University was rocked by the appointment of John F. Hatchett to head up its newly-created Martin Luther King Afro-American Student Center. Jewish organizations were especially outraged because Mr. Hatchett, a former public school teacher, had written in *Forum*, a publication of the Afro-American Teachers Association, that "We are witnessing today in New York City a phenomenon that spells death to the minds and souls of our Black Children. It is the systematic coming of age of the Jews who dominate and control the educational bureaucracy of the New York Public School System and their power-starved imitators, the Black Anglo-Saxons. . . . Our children are being educationally castrated . . . mentally poisoned by a group of educators who are actively and persistently bringing a certain self-fulfilling prophecy to its logical conclusion." Despite some outbursts of Jewish indignation, former UN Ambassador Arthur Goldberg, apparently believing Hatchett had changed his mind, gave him a clean bill of health. But the former teacher would have none of these Goldberg variations on the theme of Jewish castration, stating he meant every word he said. Later in the year, Hatchett was "fired" at full pay, after he called Hubert Humphrey, Richard Nixon, and union chieftain Albert Shanker, "racist bastards." Here the confrontation had become manifest in the highest echelons of the Academy. And from there it filtered down to the public schools that had been kept closed by a teachers' strike almost since the fall term had begun. Here, too, racial and anti-Semitic charges were exchanged. On October 23, *The New York Times* ran this two-column, two-line headline on its first page: "Racist and Anti-Semitic Charges/Strain Old Negro-Jewish Ties."

On August 15, 1968, it was reported that the Student Nonviolent Coordinating Committee was considering changing its name to Black Liberation Front. Meanwhile, its official organ, *The S.N.C.C. Newsletter*, informed us that Israel is guilty of "terror, force and massacres" in terms so harsh as to equal the crudest Arab anti-Semitism. There were the old cartoons that find their way into every anti-Semite's handbook: Israeli Defense Minister Moshe Dayan is depicted with dollar signs instead of stars on his uniform, and a dollar sign is superimposed on the star of David. We have seen these cartoons before, especially in Soviet anti-Semitic literature, and we have heard corresponding statements from official representatives of the Soviet bloc at the United Nations.

What is frightening is that the Negro—especially the so-called "militant" Negro—knows that the Jewish community can be made to pay blackmail. We find again the anti-Rothschild remarks that usually precede a pogrom, and they will lead to pogroms here unless the Jews cease their lemming flight to self-destruction, cease their masochistic penchant for alien causes. Ralph Featherstone, program director of S.N.C.C. has said: "Some might interpret what we say as anti-Semitic, but they can't deny that it is the Jews who are doing the exploiting of black people in the ghettoes. They own the little corner groceries that gouge our people in the ghettoes." (Mr. Featherstone died in March, 1970, the result of a bomb explosion in a car he was driving.)

But we are running ahead of ourselves. It is time to return to closer examination of the riots.

For a case history, I have selected Newark because the Newark riots have received a sort of post-mortem on a National Educational Television program and WCBS-TV has filmed a series called "Newark Revisited"—and because I know the city at first hand, having lived there from 1953 until 1960.

Chapter Eight
NEWARK: A CASE HISTORY

On August 7, 1967, Channel 13, the educational channel in New York, presented a ninety-minute program filmed live in Newark on the day in which the riots broke out in Detroit, the direct culmination of those in Newark. The impression given was that all "responsible" elements in the community were represented. The Mayor, the city council, some church groups, almost all local Negro organizations, the chamber of commerce, all had representatives there. Private citizens (mostly Negro) were the "extras," a kind of Greek chorus that booed and applauded the speakers. The tenor of the discussion, an attempt at "meaningful dialogue," was that many factors were to blame: the politicians, the city administration, the state, the Federal government, poverty, rats, the schools, and, above all, the police—the brutal police most of all.

But, although even a looter and a rioter participated in the discussion, no one at all represented the point of view of the "lootees." The men who had their shops burned, looted, and vandalized had no part in the discussion. They were indeed the forgotten men in the dialogue. Pleas were made for compensation to the families of those who had been killed in the rioting; pleas were made for the looters who had taken only, as one Negro lady put it, "what they needed"; even a modest and rather pathetic defense of the police was put forth by the Mayor's representative. But no one had a word to say for the unfortunate victims who had lost only their property—and their sense of security.

The producers at National Educational Television tell me that they did invite representatives from the Springfield Avenue Merchants Association but that none showed up. In later conversation with some of these merchants I found their bitterness so great that it left little patience for dialogue with "bums," as they called them, who only a few days before had been shouting "Get Goldberg!" as they ravaged business establishments without care whether anyone was in them or not. One such merchant told me that, when the mob ran toward Bergen Street, he was reminded of the tales of pogroms his father used to tell him when he was still alive (*Olav Hasholem,* he added).

Another said: "What's the use talking to them? They want what we got, only they think we stole it."

"I got nothing to say," announced another, who ushered me out of his store on Springfield Avenue. On Clinton Avenue, a shopkeeper started to reminisce about the days when he had moved to Weequahic (the "gilded ghetto of the Jews in Newark," as it has been called), when everything had been so fine, so clean, so fresh, so much a dream come true after Williamsburg, Brooklyn. His opinion of what he called *schwartze* ("blacks") was perhaps understandable but still shocking. "Me who always supported civil rights, always gave when they asked, always . . . ," his voice trailed off in words that cannot be repeated here.

Suddenly there flashed across my mind a memory of the time when my wife and I and our daughter had moved to Newark; my daughter enrolled in Weequahic High School, and my wife commuted to her teaching job in New York. At first we lived on Lyons Avenue, directly opposite Beth Israel Hospital. Later we moved to Huntington Terrace, in a neighborhood where the streets are wide and lined with maple trees and all the one- and two-family houses are surrounded by hedges and lawns. After East Nineteenth Street in New York, it seemed as if we had moved into a rural suburb. And Jews lived there contentedly, 20,000 families, in an area named for an old and lost Indian tribe. Its park was rich in foliage and flowers and lakes, where at the turn of the century the Protestant gentry had engaged in all kinds of sports, including trotting, boating, fishing, and strolling in the flower gardens. Downtown there were a wonderful library and an even better museum, with a splendid collection of Americana, both still run by members of the old families, who had otherwise become almost totally displaced in the government of the city. Not one white Protestant, not one so-called "wasp," was ever invited to participate in it. Newark was divided into five parts: a Jewish section high up in the Weequahic area; an Italian area on the North Side; Irish and East European neighborhoods in the Ironbound; and a Negro area on the South Side.

The high schools sent forth their graduates to jobs in factories, department stores, insurance companies—and to colleges all over the land. When my daughter was graduated from Weequahic High School, scholarships were granted to classmates enabling them to attend such institutions of higher learning as Harvard, Yale, Williams, Mount Holyoke, Radcliffe, and Vassar. But Weequahic was the "Jewish high school." The other high schools also had scholarship students but fewer of them, at least in the "prestige" colleges. The Jewish parents were mostly small businessmen, with a sprinkling of "royalty"—the "herring king" and "kings" of furs and fashion, as well as an enormous number of doctors, dentists, lawyers, and accountants. All these men, or almost all, were of liberal persuasion and belonged to the Anti-Defamation League of B'nai B'rith, and sent money and their sons and daughters to aid "good" causes. The "best" cause of all, outside of fund-raising activity for Israel, was civil rights.

These men mourned the dead in Mississippi, the wounded in Alabama, the uneducated in Little Rock, and every slight to a Negro anywhere outside New Jersey. But in Newark things were different. At first I could not understand it. It was almost three months after I had moved there before I realized that my neighbors always said "they" when they meant Negroes. My daughter heard the word *schwartze* for the first time in her life—from friends whose parents were members of the A.D.L. and solid supporters of integration in the South. When, however, the first two Negro children were enrolled in the Peshine Avenue School a couple of blocks from where we lived, tension was instantly apparent. Suddenly we heard the hushed voices of grown people in a state of panic:

> "What are you gonna do?"
> "I don't know."
> "I'm moving—if only I can sell."
> "Who you gonna sell to?"
> "I don't know. I wouldn't like to, but what can I do?"
> "You're gonna sell to them *schwartze!*"
> "I ain't a martyr. I'm getting out."

This exchange is quoted word for word from a conversation I overheard in a stationery-luncheonette on the corner of Osborne Terrace and Renner Street.

At that time I was engaged in helping to elect to the city council former Mayor Meyer Ellenstein, who had long been retired from public office. I went into the Negro community to solicit support and even invited Adam Clayton Powell to help. As we were able to win a good deal of this support, helping at the same time to interest some young Negroes in politics, Ellenstein was elected. When I mentioned to my neighbors that I had been down on Spring Street and Washington Street and was planning a concert-rally with Mahalia Jackson, Powell, and other Negroes, they blanched visibly. They thought that I had surely gone out of my mind. Many of them still had shops in that area, where they had settled on first arriving in Newark and before moving up to High Street; they had only recently abandoned a center built for their children worth hundreds of thousands of dollars. Now they had to make do with a couple of frame houses farther south, on Chancellor Avenue, where they were sure that the black invasion was at least twenty years away.

They recalled the constant pressure on them from Negroes, who kept pushing into what they regarded as their communities. Most of all they resented the fact that, although wherever they moved into a community the property values increased, when the Negroes came real-estate values declined and the slums like those that they had left behind in Brooklyn or Manhattan soon sprouted again across the Hudson; black slums. They blamed the South for sending uneducated Negroes north into their "gilded ghetto."

With their pride and prejudices, they never did—and most of them do not now—recognize that the Negro sees them as strangers and intruders. They are proud of their community; they remind visitors that Philip Roth (who has a rather unflattering opinion of them), Leslie Fiedler, Dore Schary, are from Newark, as are many university specialists on the American frontier, the Indians, and the Negroes. And—it is necessary to repeat—all the time there was a constant clamor for civil rights. These Jews did not heed Rabbi Prinz, whom they had imported from Germany and who was soon known all over the country; he told them to stay where they were, but they did not stay. Later he would move his temple and his congregation to the almost totally unintegrated community of Livingston, New Jersey.

They did not stay put as the Italians stayed in their neighborhood and as the other minority groups stayed in the Ironbound. For whatever reason (we shall take up this question later), they left, as they had left every neighborhood they had lived in. The old, disfranchised "Wasps" thought that the Jews, when they first came down to the wide streets of Weequahic, would destroy the neighborhood because they came with their own style of life and their own shops. Soon Bergen Street and Chancellor Avenue became busy and bustling as once New York's Second Avenue had been. Now that once-famous theater-strewn and cafe-bedecked avenue is a dreary, ill-smelling, poverty-pinched street of *bodegas*, run-down movie houses, and hippie-sponsored blandishments.

But the life and commerce that Jews bring to a neighborhood are not the kinds that Negroes like or are capable of emulating, even if they wanted to, which they do not. The problem is something else.

Fears and promises mixed with a kind of jumbled enthusiasm for causes—especially if they can be identified as "progressive"—have created a sort of social and political schizophrenia in many Jewish communities throughout the land, which their members now think will be solved near the waters of the Pacific Ocean. Dozens of interviews with Jewish storekeepers who were looted or burned out reveal the desire to go west, to California. They dismiss Watts, which is beyond their interests, as they do not intend to push the Negroes—they emphasize this point—out. Instead, they hope to enlarge the older areas of Jewish migration in Los Angeles, where Jews live in peace and tranquillity, or so these businessmen think. In many of them, however, the riots have produced a kind of numbing trauma. On September 27, 1967, the *Amsterdam News* carried a two-column headline: "Stores Available After Newark Riots."

A friend who has worked as a newspaperman in Newark for a generation and who lives in the Italian community there told me: "Nobody really cared. It was a war between the Negroes and the Jews. And many Italians just sat back and laughed, saying the Jews had it coming, especially the civil-rights rabbi." Yet, although S.N.C.C.'s representative in the "dialogue" told the Italian administration that it is time for the blacks to take over, it is the Italians whom the Negroes have the most difficult time dislodging from their homes. They do not run. They stay put wherever they strike roots in the United States. They may or may not live in slums, but they dislike hearing their neighborhoods called "ghettos." And they have worked too hard for what they have to be easily pushed around. They never establish committees to defend this, that, or the other civil-rights cause; they did not send their boys to Mississippi. But the Jews did. And outstanding among those who have helped to tear down the image of the Jewish businessman in the Negro community have been reform rabbis.

Rabbi Paul H. Levenson of Temple Ohabei Shalom in Brookline, Massachusetts, has said: "The Boston Negro may know the Jew as his landlord, and his landlord may be the one who elevated rents 25 percent when the Negroes came sweeping through Roxbury. He may know the Jew as his corner grocer who takes from him faster than he can earn it. The Negroes regard Jewish merchants as very sharp traders—the sharpest. If a Jewish restaurant owner can put ham on Jewish rye and tell a Judge that he thought the Jewish rye made the ham kosher; or if Jewish butchers drag their thumbs across the scales for Jewish customers . . . how much more so would such a despicable man not hesitate to do the same to an uneducated Negro . . . ?"

The liberal rabbi's quest for social justice often leads him to say outrageous things in an effort to placate the Negro community. That such remarks may provide a kind of justification for future looting is only too apparent.

LeRoi Jones, after the riots in Newark, wrote a poem, "Black People," which appeared in Barney Rosset's *Evergreen* magazine. The poem tells that the desired automobile in Frelinghuysen, or major appliances at the large department stores—Bamberger's, Sears', Klein's, Hahnes', Chase—"and the smaller joosh enterprises" are perhaps available with no down payment, "no money never." Now, I imagine that Bamberger's, Klein's and Hahnes' can protect themselves. But those "smaller joosh enterprises" have a problem that is not diminished by remarks like those of Rabbi Richard C. Hertz of Temple Beth-El in Detroit published in *Ebony* magazine:

> To Christians, race is not just a problem—it is an opportunity to make their religious faith vital and relevant. To Jews, the Negro Revolution seems to evolve around who is buying the house on their block.

Then he adds, "I worry about Negro-Jewish tensions because I know that the image of the Jew in the Negro community is not altogether good."

Why should the image be good when rabbis talk about members of their own faith the way they do? Ultimately, it leads men like LeRoi Jones to write:

> *Look at the Liberal Spokesman for the jews clutch*
> *his throat & puke himself into eternity. . . .*

Talking of the liberal spokesman for the Jews, brings us back to the Newark "dialogue" of August 7, 1967. At one point Mr. Allen Sagner, the administrative head of Newark's Beth Israel Hospital, which was built with funds collected from Jewish families, remarked that, as a "white liberal," and a detested one he inferred, he wondered what he could do to atone for his approval of the establishment of a medical school and hospital in the crowded Negro area of the Central Ward. A Negro woman stood up to ask him, to tumultuous applause from the gathering, why there were no black interns in his hospital. He replied rather wistfully, Bring me a young Negro M.D., and I will put him on the hospital staff immediately." Silence greeted his reply.

But the discussion illustrated a point worth noting. Aside from the black power advocates, who seemed to have a majority in the audience, participants in the discussion agreed that money—vast sums of it—is the ultimate solution to all the problems confronting the "ghetto." Superintendent of Schools Franklin Titus agreed that education was "poor" because the "plant was poor." Fix up the *plant*, he seemed to be saying, and the education level, deplored by one Negro woman who stated that Negro children were at a level three and four years below that of their white counterparts, would automatically be raised. The tragedy is that this man probably believes his statement; there would be no point in reminding him that the learning process can be carried out in a hovel if the desire to learn is there and the commitment to learn is backed up by family and community discipline. Mr. Titus was pleading, as were all the whites, but they did not impress the young "militants," who jeered and mocked and who would have none of it. "You got yours, whitey," was their message. "Now it's our turn to get ours." Nobody stood up to them, and nobody challenged their ignorance and their fanaticism.

When the Newark "dialogue" was cut off the air, the cameras turned to a quieter discussion between two Federal specialists on poverty. One was John Feild (white), the other John A. Buggs (black). Both agreed that the kind of money that the Newark officials had been talking about earlier would be a mere pittance, compared to what the Federal government was prepared to offer. Whereas in Newark people spoke in comparatively modest terms of hundreds of millions, the Federal specialists spoke in terms of tens of billions of dollars. Whereas local bureaucrats were prepared to settle for something like 5,000 housing starts a year, these two were ready with programs for one million housing starts a year. Why not? "A million housing starts and all slums would soon be wiped out at the most in five years." Do they not know that in five years much of the bright new housing of today will be the slums of tomorrow?

A generation ago all the prevailing "urbanologists" sold the nation on the merits of high-rise housing for the poor. Today almost every student of the city knows that such housing merely created vertical slums, a variation of the original—or horizontal—slums. It happened for a thousand reasons, all of which were foreseen by men with wisdom and compassion but without the bureaucrat's lust for spending other people's money as the panacea for all social ills.

Although the panaceas are a bit tarnished today, the explanations usually offered for Negro turbulence and violence—"Negroes live violent lives, unavoidably," said James Baldwin—are even more so. I refer here to the explanation—offered widely on radio, on television, in political discussion, in the daily press—of "frustration." On this subject the distinguished philosopher-historian Eric Voegelin has written:

> Frustration is a concept of individual psychology; for methodological reasons it cannot be transferred to collective behavior. If aggression is caused by frustration, it does not follow that it will subside when the suppressed desires are satisfied; on the contrary, it is highly probable that with satisfaction the desires will grow. Ultimately, and most important, there are quite a number of desires which should be frustrated when the individual is not capable of sublimating them in such a way as is compatible with the value system of the community in which he lives. *Our whole system of civilization is built on the frustration of desires which would destroy it if satisfied. Journal of Politics,* May 1940; (emphasis mine).

Voegelin invoked this argument years ago in answer to those who sought to rationalize or condone the actions of the German street mobs as the result of "frustration" over the Versailles Treaty or some such thing.

Everybody wants to show good will toward the "unfortunate" minority; everybody wants to do something for the poor "colored man." But he says: "Don't call me colored or even Negro. I am black, and I don't want your pity, and I don't want you to 'slobber' over me."

Hannah Arendt once observed: "One can hardly overestimate the disastrous effect of this exaggerated good will [on the part of certain well-meaning and enlightened intellectuals of the eighteenth century in Germany] on the newly Westernized, educated Jews and the impact it had on their social and psychological position." The same can be said of the Negroes and the intellectuals today. In the latter case, it may well be as Eric Hoffer implies: that the intellectual would rather be persecuted than ignored. This is especially true for the black intellectual.

The seeds of tomorrow's resentment and rebellions, disturbances and urban dislocations, riots and lootings are being planted today by soft-hearted and soft-headed politicians, social reformers, and bureaucrats. Frustrations will rise with expectations. And almost nobody seems concerned about this problem. A story in *The New York Times* of July 27, 1967, seems to bear out this point: "Eighteen-year-old Wayne Barton, a bakery worker, summed up the area's frustration: 'Everything is broken down. But the people still aren't satisfied. The color TV sets they stole cost money to operate.' "

At the root of much of the turbulence in the nation's cities lies the fact—uncomfortable as it may be—that Negro family life is fragmented, fragile, unbalanced. Some experts point out that many—too many Negro families in our slums, more than half the families in Harlem, for instance, lack fathers at home. They seem to believe that this difficult problem will be solved when enough *money* is poured into the "ghettos." The problem is real, and money is needed. But money and good intentions will not solve the problem in a hurry. Nor will the riots disappear. They may shift from "ghetto" to campus; from the "inner city" to "Downtown," as Congressman Powell warned; but they will persist so long as more is promised for less—more money for less violence. This way nothing is solved; only the price of peace—or blackmail—is institutionalized.

Chapter Nine
IS A SLUM A GHETTO?

I think a word should be said about the term "ghetto," and the pejorative connotations it has assumed in discussions of the Negro problem or, as it is often called, the problem of the poor. I do not mean to embark on an exercise in logic chopping or semantics. Why my concern that the term is misused as a synonym for "slum" and that the original meaning has been forgotten? My purpose is twofold: to put to rest a notion among the uninformed that "ghetto" means something undesirable *per se* and to show that it can and should mean something wholesome, that *how* people live is more important than *where* they live.

What is a ghetto? We can be certain that it is not what Bayard Rustin says it is:

> A ghetto means two things. It means government pressure on you constantly, and that means resentment towards police. It means poverty. Now, rightly or wrongly, Jews collect too much rent in the ghettos. Jews have stores which are small, and for good or ill, prices are higher.

Mr. Rustin produces no evidence to bolster his modest distaste for Jewish landlords and little shopkeepers. If Mr. Rustin, who has many Jewish friends who could have enlightened him, is so uninformed about the ghetto, it can be imagined how others less favorably placed have built up the most absurd image of a phase of Jewish life that might instead serve as an exhilarating example for American Negro life.

It is thus necessary to delve a bit into the subject, for it is soon clear that misapplication of the term "ghetto" to any American Negro community almost invariably leads to a sort of denigration of the Jewish personality, as Mr. Rustin, for instance, moves easily from "police" and "poverty" to accusation of the Jewish landlord and tradesman as gougers of "ghetto" residents.

As long as Negroes keep thinking of themselves as locked in, as literally *walled* in, it will be impossible to reach them with understanding or compassion. The *Universal Jewish Encyclopedia* tells us that the term "ghetto" originally meant an area in a city set aside for the residence of Jews "and to which they were confined by law." This system prevailed throughout most of Europe for 300 years, from the fifteenth almost through the eighteenth century. This "compulsory" system was reintroduced into Poland only after the Nazi occupation. The Pale of Settlement, which limited Jewish residence to certain areas of Russia under the Czars, was not a ghetto, strictly speaking, and is not considered such by informed observers.

Jews did not always live in ghettos before the sixteenth century. That is, they were not always *forced* to live in confined communities for the first 1,500 years after their second dispersion in A.D. 70. It sometimes happened that a *restricted* community—for Jews only—was erected for the protection of the Jews, as in Speyer, Germany, in 1084. There the first informal "ghetto" was established as an act of favor by the bishop; he granted certain privileges to the Jews, among them a wall to *protect* them and an edict to *honor* their presence in his city:

> In the name of the holy and indivisible Trinity, when I, Rudiger, Bishop of Speyer, changed the town of Speyer into a city, I thought that I would add to the honor of our place by bringing in Jews. Accordingly, I located them outside of the community and habitation of the other citizens, and that they might not readily be disturbed by the insolence of the populace, I surrounded them with a wall.

The rest of this unique and significant document shows that at least one so-called "medieval mind" was morally far above many of those who made the Enlightenment 700 years later. A comparison with Voltaire and the Baron d'Holbach, for instance, reveals how the bishop's values shine in a world too often sunk in moral darkness. They stand as a beacon, unequaled before or after the Reformation.

Areas for Jews were also granted as favors in Spain until the fourteenth century. Such were the *judería* of Valencia established by James I of Aragon in 1239 and probably also those in Zaragoza. These areas were usually set up where they would *best* serve the

Jewish community—near an important church, the lord's manor, the marketplace, and, most certainly, the synagogue. This last point is important because of Jewish religious requirements against walking too far from the house of worship.

The first *restrictive* confinement of Jews occurred in Barcelona in the fourteenth century. It was not, however, until an edict of January 12, 1412, issued by Juan II of Castile, that Jews in Spain were ordered into definable quarters behind closed walls. (This edict, by the way, also applied to the Moors.) In France too *juiveries*, "Jew streets," were established. In Italy the *Judaca*, or Jewish restricted area, was known as early as the year 1400 in Turin; *Judacas* or *Judacarias* became widespread in the sixteenth century.

Although scholarship is not too definite on the point, the term "ghetto" itself was probably derived from the word *ghèto* or *gèto* ("foundry"), which came in turn from a section of the old city of Venice where cannon were made and to which the Jews were expelled in 1516.

But the confinement of Jews behind their own walls, to set them apart and to keep the mobs from their throats, received its main impetus from Pope Paul IV's bull *Cuminis absundum* issued on July 15, 1555. Its dual purpose was to counter the Counter Reformation and to expel all Jews from the papal states, except for certain cities in which they were to be confined behind closed walls. From then on, Jews in Italy could live only in cities with established ghettos, and nowhere else. Livorn (Leghorn) was an exception, perhaps because it was a great trading port and the Italians did not wish to lose trade because of the Pope's fanaticism.

In time the ghetto took on aspects of what could be described, very loosely, as a slum if a slum is any neighborhood that is overcrowded. It had originally been decreed that any area set aside for the Jews could not enlarge itself, except vertically. Yet the people chosen of God persisted in fulfilling His commandment to divide and multiply. This population growth led to much congestion —and to high-rise apartments. As the Jewish community was forbidden to expand horizontally, it devised the multiple-unit dwelling, with one rickety story placed upon the other, and thus was invented the tenement. One early illustration shows a street in the Venetian ghetto with buildings seven stories high.

Each ghetto was equipped with gates that were locked at night, opened at dawn, and sealed on Sundays and other Christian holidays. Left to themselves, the inhabitants built a culture within a culture, with academies, shops, synagogues, schools, *mikvahs* (ritual bath houses), and entertainment halls. Although the inhabitants were deprived of knowledge of the outside world, their lives were enriched by the kind of learning and scholarship that did not require settings of brick and mortar and steel. The ghetto reaffirmed, as nothing else could, the Jews' close identification with the faith that they stubbornly refused to abandon, even under duress.

Ghetto life therefore was not all darkness and meanness. The great ghetto of Prague, which became a city within a city, is testimony to the richness of ghetto life. There, at least, the Jew was safe to pursue his faith among fellow communicants in peace. The expulsion of the Jews from this ghetto by the Empress Maria Theresa was rescinded after a year because of diplomatic pressure from most European capitals. The last of the old ghettos was abolished in Rome in 1870.

Above the cemetery gates of the Prague ghetto there used to hang a shield with the inscription:

> *Reverence for antiquity*—(*Tradition*).
> *Respect for ownership*—(*Private Property*).
> *Rest for the dead*—(*Peace*).

By 1917 not a single ghetto—in the literal sense of the term—existed anywhere in the world. But in 1940 the Nazis reintroduced the formal ghetto in Lodz, and in August of the same year a half-million Jews in Warsaw—the capital of world Jewry—were ordered walled in, in preparation for the "final solution." It is the physical wall that makes the ghetto, not the accumulation of garbage in the back yard.

As Louis Werth has written, "The ghetto shows that what matters most in social life is not so much the hard facts of material existence, [but] the sentiments, the dreams and ideals of a people." To which might be added their sense of community and family, the latter above all.

part 2

The Negro Family Problem

Chapter Ten
ITS HISTORY IN THE UNITED STATES

One thing is certain about the Negro family: No single-factor explanation is possible. Surely John Slawson, former executive vice-president of the American Jewish Committee, is terribly wrong when he tells us that "the $100 billion freedom budget proposed by A. Philip Randolph, and the massive housing and job training program advocated by various civil rights organizations are *the* answer to the problem of family instability among Negroes . . ." (emphasis mine). He later qualifies his statement, adding that "such crash programs . . . are in themselves not sufficient. . . . Implying that they are, [Bayard] Rustin and others permit themselves to lose sight, at least temporarily, of the important truth that there are things which no one can do for a man except himself."

The last statement is true enough, as far as it goes, but the first statement is dangerous, because so-called "responsible" Negroes have picked it up and used it to pressure politicians seeking easy solutions. I do think it is demeaning to the Negro, especially the young Negro, to tell him that the almighty dollar is equal to dignity and to pride. But the professional friend of the Negro loves to hear Whitney Young, Jr., say things like "Pride and dignity come when you reach in your pocket and find money, not a hole." This statement is repulsive. The young militant who rebels against it—and wants nothing to do with "Whitey" Young, as I have heard him called by a CORE official—must sense the shame implicit in the statement. It is especially shameful that his unfortunate remark should be exploited in the mass media of the nation (*Time*, for instance) to make all Negro aspirations seem shoddy.

The Jewish comedian who says, "Money can't buy poverty," is telling a kind of social truth. The world's folklore is replete with evidence that poverty is not in itself shameful. But it is a tragedy of the Negro condition in the United States that white liberals (especially Jewish liberals) have sold him on the notion that poverty is somehow conducive to criminal behavior. It would be a pity if he were to think that he could buy "dignity" with money.

Nor is the pique exhibited by Negroes (who know the facts) at white statements of the simple truth, that more than half the homes in Harlem are without fathers, very realistic. Perhaps the Negro intellectual does not really want to face the problem at all. On the other hand, perhaps his fondest wish is that whites would simply stay out of the discussion altogether and leave the problem for him to solve, or not to solve. But men of good will have a right to seek viable solutions to all social problems affecting the national polity. I shall try in this chapter to give a picture of the American Negro family and to offer some tenative *approaches* toward a solution to what I think is a problem.

Some specialists, mostly white, would like to demonstrate that the Negro family is fundamentally different from the white family and that perhaps there is no need to change its patterns at all.

Melville Herskovits, the late and noted cultural anthropologist, was convinced that "It is well recognized that Negro family structure in the United States is different from the family organization of the white majority. Outstanding are its higher illegitimacy rate and the particular role played by the mother." Dr. Herskovits was one of the first to assert the African orientation of almost all American Negro cultural traits, from speech patterns to a penchant for organizing insurance companies. In his book *The Myth of the Negro Past* he set out to disprove a number of myths.

The first myth was the notion that the Negro is "childlike" in both his African and his American habitats.

The second myth was the notion that only poorer Africans were sold into slavery. In fact, upper- as well as lower-class Africans were so enslaved.

Dr. Herskovits' third myth was that widespread mixing in this country of Africans of different tribal origin led to the loss of tribal identification. On the contrary, Dr. Herskovits argues, tribal identification was not entirely obliterated by American culture. (This theory, of course, later contributed to Gunnar Myrdal's thesis in *The American Dilemma* and helps to account for some of today's forays into black nationalism by young Negroes. It was originally an outright attack on the theories of the great historian of the Negro family, E. Franklin Frazier.)

Myth number four was that tribal culture of the Africans was so weak that it simply could not survive among the more developed cultures of Europe and the Americas. Dr. Herskovits did not believe in applying value judgments to the study of cultures; as a scientist, he would have made no *moral* distinction between the Sermon on the Mount for example, and the ritual devouring of enemies during tribal ceremonies.

Dr. Herskovits goes almost completely astray when he tries to show the African origins of almost all American Negro patterns of behavior. For instance, in reference to agricultural cooperation in certain African villages, he tells us, "Cooperation in the field of economic endeavor is outstanding in Negro cultures everywhere." What a generalization for a man of so-called science. For evidence in this country, he cites the "gang labor" in turpentine camps and elsewhere in the post-Reconstruction South. But he is stretching his argument a bit too much. He also finds analogies between American Negroes' common use of insurance as a form of savings, and primitive forms of "insurance" practiced by certain tribes in West Africa. Such scholars as Cleanth Brooks in linguistics and the late E. Franklin Frazier in social history do not agree.

Dr. Frazier, himself a Negro, in his classic study of the Negro family in America, disputed Dr. Herskovits' thesis that the Negro family is "different" because of the African background. Frazier regarded slavery as the determining fact, one that made stable family relationships among Negroes almost impossible. Most present-day students of the subject adopt the Frazier thesis with few or no reservations.

Ernest W. Burgess, in the preface to Frazier's *The Negro Family*, writes, "The first prerequisite in understanding the Negro, his family life, and his problems is the recognition of the basic fact that the Negro in America is a cultural and only secondarily a biological group and that his culture with all its variations is American and a product of his life in the United States." Unfortunately, Herskovits' book was only the first of many that would have dire consequences for Negro-white relations.

Crucial to Dr. Frazier's thesis is the statement, "As regards the Negro family, there is no reliable evidence that African culture has had any influence on its development." He is mildly amused at Countee Cullen's apostrophe to a nonexistent African heritage and attributes it to "poetic license." Today when black nationalists raises battle cries of hatred based on a past that never was, we can only regret that Dr. Frazier is not here to put matters right.

Dr. Frazier quotes from Philip Bruce's *The Plantation Negro as a Freeman:* "Under the old system, the ladies of his [the plantation owner's] family often instituted Sunday schools . . . as well as general rules of good conduct." Frazier then explains that plantation responsibility had been replaced by the nonresponsibility of the proprietor or landholder for his hired hands, sharecroppers, or tenants. "The colored people generally held their marriage (if such unauthorized union may be called marriage) sacred, even while they were yet slaves." Dr. Frazier establishes the fragility of family patterns under the old slavery system, in which the husband could be separated from his wife by a slave owner who might sell one away from the other. It is central to his understanding of the Negro family that, with emancipation, the loss of roots and rapid emigration to the cities loosened even the fragile family structure that had existed in antebellum days. "The mobility of the Negro population which began as a result of the Civil War and emancipation tore the Negro from his customary familial attachments." Now what *were* his "customary" familial attachments? According to Frazier they were the maternal family organization, "a heritage from slavery, [which] has continued on a fairly large scale."

Stable family life among Negroes—and it was common even in relatively early days—was, however, observed among families who "were able to accumulate property." Sometimes this "property" included slaves. Says Frazier:

> Those families in the Negro population that have had a foundation of stable family life to build upon have constituted . . . an upper class more or less isolated from the majority of the (Negro) population. . . . Generally, these families have attempted to maintain standards of conduct and to perpetuate traditions of family life that were alien to the majority of the Negro population. Often . . . they have placed an *exaggerated* valuation upon moral conduct and cultivated puritanical restraint in opposition to the *free and uncontrolled* behavior of the larger Negro world. [Emphasis added.]

Why "exaggerated"? And how can "uncontrolled" behavior be "free"?

If it is true that the child is the "little creature of his culture," as Ruth Benedict believed, and that the family is the root of the cultural tradition, then the widespread problems resulting from absence of fathers in much of Negro family life will not be solved by laws, poverty programs, or action in the streets. They are a fact that no amount of wishing and welfare money can change. Not that even most Negro families are without husbands and fathers. But too many are, especially in the urban centers. As Daniel Moynihan points out, "Only a minority of all Negro children reach the age of eighteen having lived all their lives with both parents." Whether this pattern results from the "disorganizing effects of freedom," as Frazier put it, or from a reluctance to accept Western monogamy, it creates problems affecting Negro education, development, and assimilation into American life.

Dr. Frazier argues that illegitimacy is a consequence of city life faced by any simple peasant folk whose roots have been suddenly torn up. The simple peasant girl may be in for some surprises as she discovers "romantic sentiments" in such a song as that of the woman who complains:

I loves dat bully, he sho' looks good to me.
I always do what he want me to,
Den he don't seem satisfied.

It is not "romantic sentiment" that the forlorn "peasant" girl has discovered, but something else. It is the sophisticated, but irresponsible, attraction of the hip ofay for the dissolute black. Even Frazier seems vaguely aware of this attraction when he remarks elsewhere, "When they sing, they sing the blues which represent the conscious creations of song-writers who supply songs for more sophisticated sentiments." But if he knew the sodden sophistication of the origin of some of the blues he might have been pleased to learn of the disappearance of the "mulattoes"—the half-black gentlemen who in the past showed "considerable prejudice toward blacks with the result that they tended to form separate communities." Fact is the more white genes floating around in some Negro's blood, the more black-conscious he has become. No wonder Big Bill Broonzy's ballad is heard no more in the land:

If you's white, all right
If you's brown, stick aroun'
But if you's black—
Stand back, stand back, stand back.

But I believe that Dr. Frazier is on less solid ground when he argues that "Family desertion has been one of the inevitable consequences of the urbanization of the Negro population." The reason is that he seems to be of two minds when he speaks of a kindred subject like "illegitimacy." He is convinced that "poverty and city slums"—by the way, the word "ghetto" appears nowhere in his book—contribute to widespread illegitimacy among Negroes. Yet he is just as sure, fifty pages later, that "Widespread illegitimacy can still be found in the rural communities." He adds that this kind of illegitimacy is "a harmless affair," for "it does not disrupt family organization and involves no violation of the mores." Mr. Frazier does not say which mores. And why it does not "violate" them is also unclear. Dr. Frazier quotes from the great American founder of sociology, William Graham Sumner, who wrote in *Folkways:* "So long as customs are simple, naïve, and unconscious, they do not produce evil in character, no matter what they are. If reflection is awakened and the mores cannot satisfy it, then doubt arises; individual character will then be corrupted and the society will degenerate." This true and terrible statement is unfortunately misunderstood by Dr. Frazier. Sumner is stating a universal fact of Western civilization. His comment is not a warrant for permissiveness. To use it to bolster the conception of the "naïve and ignorant peasant folk" who come to bad ends in the cruel city is completely to misunderstand one of this country's greatest thinkers.

Frazier offered his solution to the problem posed by "the waste of human life, the immorality, delinquency, desertions, and broken homes which have been involved with development of the Negro family in the United States," calling for "large-scale modern housing facilities . . . for the masses of the Negro population in the cities." Thirty years ago Frazier was not the only one who shared this delusion. We know better now, and some of us knew it even then. But it was a myth-ideal and, like so many ideals of the late 1930s this one has gone up in the flames of riot-torn cities as the "large-scale housing facilities" have become larger and larger and the slums uglier and uglier.

Was this development inevitable? Perhaps. But we have a clue suggesting that it was not. From a study by Jessie Bernard we learn that by the second decade of the twentieth century Negro infants born in wedlock "of a stable relationship" had reached the unusually high proportion of 89 percent. Then the figure began to decline. "Some time after 1917," she wrote in *Marriage and Family Among American Negroes*, "the forces toward institutionalization declined, or were counteracted by other trends." What these other "trends" were is not clear. The only description offered is that "the downward phase of the trajectory involved nonconformity to both the institutional norms imposed by society: life-long commitment between the two parties, and their support and protection until maturity of children born in union." It is not very satisfactory, but it is all she offers. Surely the facts are known, but the causes are unclear from the sociological jargonizing that follows. For instance, Dr. Bernard tells us that for some Negroes "conformity to the institutional forms of monogamic marriage was only superficial . . . [because] they lacked traditional preparation for such a union, the duty-responsibility-commitment complex which constitutes monogamic marriage remained foreign to them." This statement is followed by insistence that *most* Negroes today live in a "duty-responsibility-commitment complex." Then comes the kind of patronizing remark that is common among even well-meaning scholars in the field: "They remained innocently unconscious both had entered into a mutual pledge. . . . The index of the institutionalization of marriage . . . reveals a marked change in marriage and family among Negroes since World War II and especially in the 1950's and 1960's"—or, as has been suggested, since the years of *massive relief*. No convincing argument has been put forward to suggest any other reason for the continued rise of the matriarchal family in the last three or four decades. Everybody has observed the phenomenon, yet so far nobody has come up with a more satisfactory explanation. Explanation is not, however, the most important goal; some way of treating the phenomenon is.

It does not do to "cop out" in the face of hard facts. After describing the upward spiral and the wide prevalence of the "female-headed" family among Negroes, Dr. Bernard is quite sure that "One must dismiss, of course, any racial interpretation of the increase in female-headed families," Why one *must* dismiss such an interpretation is not explained except by the following *non sequitur:* "This was neither the *typical* Negro family nor a characteristically *Negro* family even in 1960. Furthermore, the racial factor has remained constant and cannot, therefore, be invoked to explain this change."

In a careful study, "Marital Instability by Race, Sex, Education, and Occupation Using 1960 Census Data," J. Richard Udrey states, "The far greater instability of non-white marriages is shown *not* to be attributed solely to the general low educational and occupational status of this group, *but a characteristic of non-white groups of all educational and occupational levels*" (emphasis mine). By "non-white" he means Negroes mostly.

The fact is that after Emanicipation and Reconstruction the Negro family did tend to show patterns of stability. Dr. Bernard's information on this point is valuable. Although the widespread absence of the male parent as head of the family had been noted and usually attributed to slavery, it was believed that with freedom and the Negro's entry into the community this pattern would straighten itself out in time. But with the Negro family a sort of Gresham's law seemed to prevail, with unstable families "driving out" stable ones. Why has never been adequately explained. But such a plethora of theories, analogies, and generalizations has been intruded into the discussion, culminating in the controversial *The Negro Family: The Case for National Action* by Daniel P. Moynihan, that it finally elicited from James Farmer this angry retort: "We are sick unto death of being analyzed, mesmerized, bought, sold and slobbered over." I sympathize with Mr. Farmer.

Kenneth Clark, a Negro psychologist and perhaps the soberest of all Negro spokesmen finds Moynihan's arguments that vast sums of money are necessary to halt the flight of male Negroes from their families—degrading to Negro dignity. It is worthwhile noting that the Moynihan thesis has found most of its support, some of it for very sound reasons, among white men. But the soundness of that support does not erase the fact that Negro leaders—almost to a man—from right to left, from moderate to militant, from integrationist to segregationist—have all rejected it. Why?

Because much heat has been engendered by the Moynihan Report it is worth summarizing its outstanding conclusions, as well as its repercussions. Perhaps some suggestions offered in the process will apppear attractive to Negroes. In this area as in many others, only the Negro community can solve its own problems, but it must first recognize them as problems and then decide to do something about them. I do not suggest that Negroes should choose one or another solution. I emphasize that the Negro "problem" is for the Negro to solve; the liberal notion that the Negro problem is also a white problem, a notion popular until recently, is simply not true. Regardless of first causes, of who was guilty for this or that situation that resulted in this or that evil, the fact is that no amount of guilt or self-righteousness will aid in solution of the problem.

Daniel Moynihan is disturbed because the Negro community refused to respond with hosannas to President Lyndon B. Johnson's speech "before an audience of 14,000 persons on hand for the graduating exercises" at Howard University, as he tells us in a well-argued defense of his views in *Commentary* in February 1967. What he does not say is that his was the chief hand in writing the speech, which was perhaps part of the reason for its failure. The same speech, spoken by another President, perhaps by John F. Kennedy, would probably have received better notices than it did from the lips of President Johnson. The words simply did not suit the man. Not that the President did not believe everything he said. He simply did not *sound* convincing when he said that Negroes in the United States, despite all progress recorded in the previous dozen years, had not reached the end, "not even the beginning of the end . . . perhaps the end of the beginning." Johnson simply does not talk that way. It is alien language, smacking of the academy, of the café, of the sophisticated jargon fashionable at Harvard University but not at the Pedernales Ranch in Texas. He went on in un-Texas fashion about not doing enough and claimed that "Negro poverty is not white poverty." By "racializing" the problems of the poor in this way, he seemed to startle his almost all-Negro audience. As Mr. Moynihan tells it, "The President proposed 'no single easy answer.' Some recommendations were obvious enough: jobs to enable men to support their families, decent housing, a welfare program better designed to hold families together, health, compassion." Then the President announced that he would propose a White House conference of scholars and experts. "Its theme would be," Mr. Moynihan informs us, " 'To Fulfill These Rights,' a phrase echoing the great assertion of the Declaration of Independence. And he dedicated his administration to this epic undertaking":

> To move beyond opportunity to achievement; To shatter forever not only the barriers of law and public practice, but the walls which bound the condition of man to the

> color of his skin. This is the next and more profound stage of the battle of civil rights. We seek not just freedom of opportunity—not just legal equity but human ability—not just equality as a right and a theory, but equality as a result.

I must confess that I do not know what the quest for "human ability" means. How does one go about implementing such a quest? Shall we pass a law that everybody has a right to "human ability"? Or were the words just rhetoric, a kind of rhetoric that not even a Solomon could interpret?

Johnson apparently expected to be hailed as another Lincoln after this speech. But when it was over the audience sat back "stunned in silence . . . finally they applauded out of shock and self-identification," as one reporter put it. Maybe. Mr. Moynihan, however, did not expect the vengeful criticism that descended upon him. The President seemed relieved that no one was really holding him responsible for these statements and let it be leaked that they were all the thinking of Daniel Patrick Moynihan.

"It was a bold beginning," Mr. Moynihan summed up the President's speech. "Yet before half-a-year had passed the initiative was in ruins, and after a year-and-a-half it is settled that nothing whatever came of it." Nothing indeed? Well, something. For it roused the Negro leadership to anger as no other public statement in recent years had done. This reaction was perhaps the most disappointing of all, for Mr. Moynihan is one of the best-informed and best-intentioned men writing on Negro problems today. Yet the entire Negro leadership was upset. Before I try to suggest why, it is well to let Mr. Moynihan have the floor so that his proposals may be understood objectively. Here is what he had argued in the Report—and what he presumably still believes:

It is necessary for Congress to come up with a "national family policy," which would supplement traditional concern for individual rights.

> The report on the Negro family was intended to demonstrate its relevance and thereby to persuade the government that public policy must now concern itself with issues beyond the frame of individualistic political thinking. The second objective was connected with and flowed from the first. Family is not a subject Americans tend to consider appropriate as an area of public policy. Family matters are

> private. For that very reason, to raise the subject in terms of public policy is to arouse immediate interest; edged with apprehension, but interest nevertheless. That was the simple purpose of the report: to win the attention of those in power.

What Mr. Moynihan proposes—and still proposes—is that every family in the land receive a Federal allowance for all children under the age of eighteen, so that the stigma will be lifted from Negroes as the primary group presently receiving aid from the Federally funded program of Aid to Dependent Children.

He has since been more successful with a Republican administration than two previous Democratic regimes, although Mr. Moynihan is, himself, a rather liberal Democrat. In any case the Nixon plan for a $1600 base for all poor families of four in the land—on or off relief—is designed primarily to provide incentives for the male head of the Negro family—the father—to stay home, and suffer no penalties for trying to supplement his relief allowance by outside work. But the Right has found it too much, the Left too little, and the vast Center—Nixon's "silent majority"—finds it simply repulsive. For them it is still relief, a handout, something given to those they consider to be "shiftless" and "lazy." They fail to understand the subtleties in the program—and care even less.

To support his point of view, Mr. Moynihan cited from Dr. Frazier's classic study of the Negro family, incorporating much that was valuable in the book. Actually, Dr. Frazier was quite optimistic about the future of the Negro in this country. The last lines of *The Negro Family* (published in 1939) read: "The gains in civilization which result from participation in the white world will in the future as in the past be transmitted to future generations through the family." Dr. Frazier's ideas, unfortunately, were almost totally ignored by Negro leaders. But that is a Negro frailty observable for generations. The Negroes—more than any other people in the land—have suffered from bad leadership. How to resolve the problem is something my old friend Claude McKay, the Negro poet, worried about till the day he died. "Every black demagogue," he argued, "rouses our passions and, unlike the whites, where only some demagogues are capable of rousing your interests, we respond *only* to demagogues—it seems."

Dr. Frazier concluded his work with a statement of hope that

Negroes would build a better life for themselves *on the basis of the family*—for all civilization seemed based on this relationship. But in this area the Negro has fared worst. Every program designed to aid the "poor"—by 1968 the term had become a synonym for Negro—has turned out to be destructive of stable family life, regardless of the number of rats killed and amounts of welfare granted. Mr. Moynihan, borrowing from Dr. Frazier and others and believing that without the institution of the family all programs to aid black men will amount to nothing, did at least try to approach the problem in all its immensity. He thought he had a solution, but it turned out to be no solution at all, for it was based on a false premise only hinted at in Frazier but elaborated by Moynihan and since exalted by certain Negro leaders.

I refer to the concept of a "displaced peasantry"—the notion that peasants who come to the big city with few skills and different folkways find their cultural patterns, particularly of family life, disrupted. (I doubt that the southern Negro fits this description at all, but as this book is not a work on economic theory, I shall let it go.) This theory is a blatant misstatement of the typical behavior of peasants and of their family habits, yet it has been passed over by most critics as if it warranted no retort or even investigation.

Surely if the argument had any merit one could cite comparative studies of breakdowns in family structure and life and of male irresponsibility among uprooted peasant families from abroad. But there are no such studies, at least none that bolster the Frazier-Moynihan thesis. It should be borne in mind that with Dr. Frazier—a careful scholar—the thesis was largely a matter of speculation. But with Mr. Moynihan it has become an assumption. He is essentially a reformer, not a scholar. Perhaps he does not really believe that the breakdown in Negro family life is similar to that of other groups transplanted to urban "industrial" centers like Copenhagen, Dublin, and Glasgow. Dubliners will be startled to hear that their city is an "industrial" center. But residents of Copenhagen may not be pleased to learn that their family structure shows disintegration patterns and absence of fathers like those of Negro families in Harlem; the claim is not true. Danish peasants did not suffer family disintegration when they moved into the urban centers; neither did Irish peasants when they went to Dublin or to the United States. As for the canny Scots, not in Glasgow or in Loch Lomond did their families show the kind of breakdown shown by the Negro family in this country. Mr. Moynihan remarks without any substantiation in facts or figures: "The kind of female-headed family now so common in Negro slums is nothing new. It has been and remains a commonplace feature of lower class life in industrial society. The Negro experience may be a particularly intensive one, but San Juan and Copenhagen, Glasgow and Dublin have or have had their counterparts." He is talking pure fantasy and not very revealing fantasy at that. No wonder he outraged the sensibilities of Negro intellectual Ralph Ellison: "Moynihan looked at a fatherless family and interpreted it not in the context of Negro cultural patterns, but in a white cultural pattern. He wasn't looking at the accommodations Negroes have worked out in dealing with fatherless families. . . . How children grow up is a cultural, not a statistical pattern." Ellison is right, of course, to the extent that he is willing to accept the notion—not original with him—that "Negro" means "different" and that Negro life patterns differ *essentially* and *qualitatively* from white patterns. More than fifty years ago H. W. Odun wrote in *Social and Mental Traits of the Negro:*

> Let the influence upon the Negro child, at least so far as the school is able, to effect this end, lead him toward the unquestioning acceptance of the fact that his is a different race from the white, and properly so; that it is not necessarily a credit to imitate the life of the white man.

The goal may not be the same as Mr. Ellison's goal, but the direction is the same. Certain young blacks who now call for an *apartheid* of their own may have recognized that competing in achievement with the white man is impossible. The Moynihan Report received unkind treatment from the left and from almost every Negro leader, left, right, or center. Yet it was the first *serious* attempt to come to grips with the fundamental problem affecting general black-white relations in the nation: the family. In his report, Moynihan paraphrased F. Scott Fitzgerald's felicitous observation "The difference between the rich and the poor is that the rich have more money": "We seem to be unable to recognize that to be poor is not to have enough money." What is *enough* money? And for whom is it enough? Actually what is enough for Mr. Moynihan would be abject poverty for his friend the late Senator Kennedy. Where the family is concerned, money is perhaps an insignificant factor. A zillion dollars will not alter effectively rooted habits or customs. Possibly only force, as applied in Red China, can impose a new "viability" in place of established traditions. We are not yet sure even that can be done.

But the "buck mentality" of Whitney Young, Bayard Rustin, and the leftist liberals—the notion that money is a cure for everything that is wrong with society—*is* a great hindrance to serious thought. "The West believes that man's destiny is prosperity and abundance of goods. So does the Politburo," Whittaker Chambers once said. Some people who know little of economics and less about money think that more of the latter will solve problems posed by the former, without realizing that money is not merely pieces of green paper that a stingy government hands out or withholds at will. Many people on the left think this way, Mr. Moynihan is essentially one of them. His ambivalence is worthy of Hamlet. "The nation needs the liberal Left," he stated in defense of his Report, yet he also states that "the liberal Left can be as rigid and destructive as any force in American life." Why the nation should "need" such a "destructive" force is not further elucidated, but Mr. Moynihan is not quite ready to part company with it. That is a pity, for he is one of the few writers in public life who are only occasionally guilty of the linguistic monstrosities with which sociological writing abounds. Words like "underemployed" and "overemployed" are there, but they do not stick in one's eye.

His concern with the problem of the disintegrating Negro family is genuine, as we have indicated. But his proposed solutions are unconvincing. He does suggest that, in the end, the Negroes must solve the problem for themselves. But on this problem no white man dare speak, without being "clawed" by his critics. That is what happened to Mr. Moynihan. The greatest blow was delivered by spokesmen of the Jewish academy, from the Ferkauf Graduate School of Education at Yeshiva University, to be exact. Their criticism was so biased and uninformed as to warrant Mr. Moynihan's remark that it smacked of the "scholarship of Che Guevara." But Mr. Moynihan's defense of his position reveals that he can no more offer a valid solution than can any other middle-class liberal.

All he can suggest is that, if the government will guarantee a fixed annual income to all lower-class families and provide $12 a month for each dependent child, the Negro Family will stop disintegrating and fathers will be content to stay at home minding their manners, their children, and their wives. Solving the problem is *not* that easy, and I do not suggest that Mr. Moynihan thinks it is. But he defends himself petulantly and is too often ready to grant points to his opponents. He did not say some lower-class youths lack "employability"; he said that they lacked "opportunities"; and so on. Chances are that they lack both. So where do we go from here?

Mr. Moynihan declares, "The time when white men whatever their motives, could tell Negroes what was good for them, is now [February 1967] definitely and decidedly over." Well, not *absolutely* definitely and decidedly; Mr. Moynihan is still at it, and he has much to say that is cogent. But he is afraid that what he says will be misconstrued by white liberals—especially Jewish liberals like Arther Lipow of the Jewish Labor Committee. The Committee, by the way, is about as necessary as the Jewish Freeland League, an organization founded more than fifty years ago to *seek* a homeland for the Jews and that is apparently still seeking it, overlooking the fact that the State of Israel has just celebrated its twentieth birthday. Mr. Lipow, according to Mr. Moynihan, who apparently takes such things seriously, denounced his Report as "distorted . . . disgraceful . . . ideological rationalization . . . ," and so on. In an angry retort, Mr. Moynihan expresses worry, lest his findings be misconstrued as seeming to "invite the charge of raising the same old canard of innate differences between the races." He assumes that his readers will understand. I do not understand. Many distinguished men have "raised" the question, among them Franz Boas and Nobel Prize-winner William Shockley. James Bryant Conant, formerly President of Harvard University, has *raised* it, and W. E. Burgess, who wrote the introduction to Frazier's book on the Negro family has underscored the "qualitative" differences between the Negro family and the white family. For truly open minds there are no closed subjects, not even in this area of national controversy. The noted Jewish publicist Menahim Boraisha said of the Jews, "We are different with the memories of thousands of years of being different."

But too many academics believe that with socioeconomic equality such cultural differences will disappear. Ashley Montagu, an anthropologist at Rutgers University, writes: "It will be generally agreed that those persons who are readily identifiable as Jews almost always originate from the lower socio-economic classes of their community." What he means is that if you raise Jews from "lower-class" status nobody—not even their best friends—will recognize them as Jews. So be it. Mr. Moynihan is influential in perpetuating belief in socio-economic solutions, for he puts too much reliance on money to help solve the terribly difficult problem of the Negro family. Actually, Mr. Moynihan is afraid of the implications of his thesis. What if all the money in the national treasury does not solve the problem but only exacerbates it, as is quite possible? After all, in Detroit and New Haven, where *more* money and accommodations were provided, some of the *worst* riots took place.

Chapter Eleven

MONEY CAN'T BUY POVERTY

Irving Kristol, a keen observer of the sociological scene, has written that, although "it used to be thought that the 'undeveloped countries needed nothing but money to catapult them into modernity,' it is now clear that what goes on in the conscious and subconscious minds of the peoples of these countries is far more important than what goes into their pockets. . . ." Furthermore, Mr. Kristol asks, "How else explain that in Latin America, the immigrant—whether German, Italian, Japanese or Jews—usually manages to achieve a varying affluence, while the indigenous population seems incapable of doing so. . . . A host of investigators are now wrestling with the mystery of the 'achieving minority.'" A "host of investigators" but not Mr. Moynihan, not Bayard Rustin, not a host of others.

The New Haven experience at the end of August 1967, which included rioting, looting, and fire-bombing, and the introduction of the new nerve gas Mace into our hot summers was indeed disheartening to pedagogues who had taught that money would provide riot insurance for the "central city," the new synonym for the urban slums.

New Haven had been the envy of every other city in the land. A liberal mayor, Richard Lee, had not only been able to corral more money for his town—$120 million in Federal urban-renewal funds, or $800 for every man, woman, and child in the city, plus $60 million in state and local funds and $250 million in private construction; he had also started early and had hired the big "brain truster of poverty," Mitchell Sviridoff, to start the program. So successful had Mr. Sviridoff been in New Haven that he was hired away by Mayor John V. Lindsay to repeat his performance in the more affluent metropolis of New York. (He did not last long in New York.)

Mayor Lee's faith in urban renewal, job training, and education was genuine. (We may dismiss Mr. Sviridoff's post-mortem on the riots. He stated that he had run out of "luck.") The trouble is it was not enough, and the New Haven riots came as the greatest shock to the element in the nation that believed that the young Negroes' main goals were jobs and education, especially a high-school diploma. Even as the riots were blazing in New Haven the Senate Judiciary Committee was hearing from Dr. Nathan Cohen of the University of California, Los Angeles, (U.C.L.A.), who had just completed an exhaustive investigation of the Watts riots that revealed that most of the rioters there did indeed *have* high-school diplomas. This testimony led Senator Sam Ervin of North Carolina to remark drily that today's Negroes have more educational opportunities than Abraham Lincoln did. The real answer to Dr. Cohen, Mayor Lee, and all the seekers of "instant freedom" for the black man was, however, given the same day by Willie Ricks, the original coiner of the "black power" slogan. Mr. Ricks said of Dr. Martin Luther King's call for "massive disobedience" that he hoped to be able to turn this into "another massive Detroit."

As every measure has so far failed to quell the tide of black discontent, it is becoming clear that, until or unless a great many Negroes are prepared to engage in middle-class commerce or manufacture in their own neighborhoods, nothing much will come of efforts to help lift the Negro up by his strapless boots. (It is a favorite remark of Bayard Rustin's that the Negro in America has no straps on his boots with which to lift himself up.) The point is that, until they do so on a massive scale, they will simply have to be satisfied to remain the best-paid, best-housed, and best-fed colored population anywhere in the world, better off—by and large—than most white people around the world. No society can solve the problems of the "less than one percent." It is enough if it can solve them for 90 percent. It is still only the totalitarian states that propose to solve all their problems; they have solved their unemployment problems only by placing the entire population in the equivalence of slave-labor camps. Money is not everything; the comedian knows what he is talking about when he says that money cannot buy poverty. He is closer to the truth than the sociologist who roils the social scene with pat solutions that sound no less absurd than when the same comedian says, "Papa buy me a slum." When papa demurs, his wife coaxes him along by saying, "All right, buy him a slum, but make sure it is in a good neighborhood." Sigmund Freud would have understood that there is more social truth in these gags than in some volumes of what passes for social thinking in the United States these days.

Bayard Rustin argues that the Negro would be doing about as well as any other immigrant who arrived in one of our big cities at the turn of the century except that nobody is offering to buy his muscle power fast enough. He elaborates by arguing that, when the immigrant—especially the Jewish immigrant—walked off the boat, he brought his muscle power with him; because muscle power was in great demand at the time. "In five minutes" (I have actually heard him make this claim a number of times on radio and television) he was recruited by a boss waiting on the dock. If only Mr. Rustin knew the real facts! I can hear a sigh go up from hundreds of thousands of Jewish immigrants who arrived without muscles, without homes, without knowing the language, without jobs or promises of jobs, but with hope that here, free, if only given a chance and not pogroms, they would "make out." And they did. Nobody waited for them at the boats to give them jobs, although many thousands of them did go to work in sweaty little shops, sometimes earning as much (or as little) as the bosses; many tens of thousands more took baskets and peddled needles and pins, bagels, borscht or rented pushcarts and sold pickles and pretzels or halvah and herring. They lived in crowded conditions in rickety tenements like the one my family lived in on East Third Street in New York City. There were three rooms for seven people, Two adults and five children. There was one iron sink with a single faucet giving cold water. There was no bathtub, and the toilet was in the hall. But we never went to school dirty, and no one served us hot lunches in school or looked out for our special interests; nor did we have a Jewish coordinator. In our schools, with toilets in the yard, Irish teachers with brazen red chests and brass-tongued rulers in their hands hammered into us the learning that we went there for: reading, writing, and arithmetic. Woe betide us if we brought home anything less than A for conduct, although B might be tolerated in a scholarly subject. Friday and Saturday, at least, we ate well, for on those days Mother took in boarders to help pay the grocery bills.

We grew up to become doctors, lawyers, artists, writers, senators, and Supreme Court justices, no matter what it cost our parents. We overcame our circumstances; most of us resented poverty (when we were old enough to make the distinction between being poor and living in poverty), and we triumphed over our environment despite all material obstacles in a society that said, in effect, we shall do nothing to impede your movement away from the slums, but neither shall we do anything to help. I recall that I personally had reason to be "frustrated" because I saw how the "other half" lived while working for a laundry service. I was hired to ride in a one-horse wagon picking up and delivering laundry in Greenwich Village, especially in Washington Mews, where the houses were beautiful. I was thrilled that people could live that way, and I yearned more than anything else to live in the Mews. When I grew older, I would always, when walking with my wife and baby daughter from our East Nineteenth Street apartment to Washington Square Park, detour through the Mews, still hoping that someday I would live there. I never made it.

Although resentment may have turned many poverty-stricken Jews into argumentative young radicals and later the Depression may have made them more radical still, it never had the effect that it apparently had on some heiresses to Jewish fortunes. Many of them dedicated themselves to kinds of fanaticism that we passed by; most of us who became Stalinists, Lovestonites, Trostskyites, and even young socialists outgrew these theories as the nation grew up with us.

I also recall how one day a boarder brought home a small violin and told me to play it; by luck I plucked out something that my parents were sure they recognized as a tune, and a violin teacher was sought. Not finding one for private lessons, my parents discovered on East Third Street, where we still lived (though a bit farther east in four rooms with two faucets on the sink—one for hot water, one for cold—and a wooden tub in the basement that we never used), a place that I had never known existed, even though it was only a couple of blocks west of my home. I refer here to the Music School Settlement, an oasis on the East Side. Where my immigrant parents found the quarters to pay for my lessons I shall never know. The place gave off a strange odor as we entered to register and was staffed by people the likes of whom I had never seen anywhere in my young life—all clean but not like the Irish teachers on East Fourth Street. An aroma of strong coffee filled my nostrils, and I remember it to this day; it was so unlike the familiar *café au lait au juif* (made of poor coffee grounds, chicory, and boiled milk; I still shudder at the thought of it). I was made welcome in the red-brick Georgian house where the Settlement was housed, although it was *not* Jewish-supported. The music school was headed by Melzar Chafee, who we all believed was of American Indian stock, with his white hair, clipped gray mustache, and ruddy cheeks. He instilled in us a respect for the violin and for Baroque and classical music. It was not until I was fairly far along in my musical education that I heard any composer later than Beethoven. We studied and learned; we listened to the works of Vivaldi, Corelli, Tartini, Bach, Mozart, and only occasionally Beethoven. We played in the string orchestra in a beautiful auditorium. There was a magnificent library with leather chairs that made no concessions to our height and books that made no concessions to our minds; the books were not adulterated volumes on Mozart for little boys but adult biographies and stories of music and musicians. There was also an open-air gym where we played basketball on paved stones, which skinned our knees; it had a shower room that was a boon, as I never took a bath in our wooden tub. As I grew too tall for my half-size violin, my teacher of blessed memory, Miss Evelyn Mellon, gave me her violin, which had been a gift to her from a Connecticut fiddle maker. It was a magnificent instrument, and the day that she gave it to me she took me to hear the Flonzaley Quartet and introduced me to the musicians. I cherished the fiddle and played with Miss Mellon in the high-ceilinged lesson room that was like a monk's cell, with a pane of glass high in the door so that students could not look out but Mr. Chafee could look in, for he observed every pupil's progress. We were there not as charity cases; no one studied for free, but the cost of the lessons could be reduced to an absolute minimum on the philosophy that what one obtained for nothing was worth exactly that. I studied with Miss Mellon in her little room at the school, where during my Friday-evening lesson she would serve me coffee, the delicious aroma which I had first inhaled on entering the settlement house. She would wipe the grimy sweat from my face with a silk scarf, which gave off its own perfumes; the charm, elegance, taste, and all-around wonder of the world she introduced me to have never ceased to arouse awe in me. I was often observed, as I have said, by Mr. Chafee. He apparently noted that I needed brushing up in my bowing technique, so a specialist in that field was assigned to me, a tall, beautiful young lady from New England. But, when she took hold of my shoulders and held my right arm up as she bent over to demonstrate just how deftly I should hold my bow, I was aware not of fiddle or bow, or Vivaldi or Tartini, of Miss Mellon or Mr. Chafee, but only of this lady's firm white bosom and her intoxicating perfumes; and I began to sweat even more than usual, and she too would wipe my brow. I knew—I was twelve then—that I would either have to give up the fiddle or stop studying with this lady. I asked and was granted permission to study with Harold Berkeley, a rather well-known concert violinist at that time; he smacked my knuckles with his bow after the first lesson, and I begged to return to Miss Mellon, with whom I continued until one day (I must have been about fifteen) when Mr. Chafee asked me to play for him. He looked at my violin, saw it full of resin dust, wiped it slowly as he stared at me with contempt and said "Play." I started on a rondeau by Beethoven that I knew I played well because Miss Mellon said I did, but he stopped me almost immediately, saying that he knew all about that. He pulled out some Kreutzer exercises and told me to play them. The sweat started up on my neck, and I had played no more than a minute or two when he closed the book, stopped me, and told me not to waste the time of the school any longer. He departed, and I knew that I had disappointed him more than he had shocked me.

At that time I was more interested in the club activities and the summer vacation to Newfoundland, New Jersey, which the settlement house provided for a couple of dollars a week. I spent summers of carefree joy in an environment as beautiful as the nation provides, especially near Green Pond Lake. The house we stayed in was spacious, but we had the choice of living by ourselves in tents or in a big barn of a dormitory. The fee for staying at the camp was predicated on the fact that we would practice our instruments, which were mostly violins, cellos or the piano. I have since returned to Newfoundland to try to recover exactly what we had there. Once when my friend Oscar and I "ran away" during a long Thanksgiving weekend and found ourselves in Newfoundland, we went looking for the shuttered house. A man and his family recognized us and invited us in for Thanksgiving dinner. It was the very first time in my life that I tasted homemade pumpkin pie, which I simply could not swallow; I fed it to the dog at my feet, whereupon my host, seeing my plate so soon empty, gave me another large slab, which made me ill. I was glad to return to good homemade *kashe varnishkes*, which would have made *him* ill, I am sure. But I have never forgotten the school where I received my early introduction into Anglo-Saxon or old American culture and where my mother never intruded herself as a member of a parent-teacher organization to tell Mr. Chafee, Miss Kibbe the librarian or the head of the violin department how to conduct the musical education of a bunch of wild and sometimes unhouse-broken boys from the East Side of New York (and from West End Avenue too, for the school had not yet acquired the snobbishness of the professional helper of the poor). Nor would she have welcomed my violin teacher's advice on how to make cheese blintzes or gefilte fish. The teachers were professionals, in the best sense of the term, and so was my mother as homemaker.

When my daughter started on the piano we sent her to the same music school on East Third Street; of the staff I remembered only Miss Kibbe the librarian was still there. There were notices on the bulletin board announcing the meeting of the parents' association, the outing, the fund-raising campaign for a prominent charity; also there were squalor and noise in the hallways. I went up to Miss Kibbe, and she remembered me. I asked what had transpired to turn the oasis on East Third Street into a shrieking bedlam, into just another settlement house for the poor. She shrugged, and her eyes took on a faraway look, as a youngster shouted up at her, "When we gonna go on the picnic?" I walked away and left my daughter behind, knowing that she would not last long as a piano student in the teeming "ghetto," as it would be described today by the social workers. Although there was still some musical discipline left, it was not the same old school, or perhaps it only seemed different.

A neighborhood may be crowded with tenements, and people may be poor without living in a "culture of poverty" or even considering themselves "deprived." We knew it then instinctively, only to have it rooted out of our consciousness by our own sons, who went to universities where they picked up wisdom alien to the Jewish tradition, that *poverty is a crime*. In my youth to equate being poor with being criminal would have elicited a smack across the face of even the brightest fledgling scholar from Harvard. The Jewish boys of that generation were not going to Harvard to learn; they were going to City College, and their sisters were attending Hunter College. Later some of *their* sons and daughters became professors at Harvard and Mount Holyoke College and sent *their* sons and daughters to these institutions of traditional Yankee learning.

The East Side that I remember is still there. The Yiddish signs have been largely replaced by Spanish ones, but the owners of the dress shops and shoe stores, if not of the *bodegas,* are still Jewish, holding on till time—or the next riot-inspired fire—consumes their little businesses. Now the area teems with welfare agencies—local, state, and Federal—all joined in what they believe is altruism, but that will quite likely diminish the dignity of the community. I recall a conversation between two old-timers in Tompkins Square Park, where the hippies now hold sway. Sam was reminiscing about old times with his fellow benchman Morris. They inquired about the children (*baruch hashem*!) and congratulated each other on the doctors, lawyers, and professors their sons had turned out to be and on the handsome marriages their daughters had made; then Sam suddenly said, "Ain't it a pity, Morris, they don't have the advantages we had?"

"Advantages?" replied Morris.

"Yeah, advantages. *We* didn't have a relief bureau around the corner when *we* arrived."

"That's a good joke," said Morris, "I'll have to tell it to my son-in-law Marvin the professor the next time he comes for dinner."

This kind of humor exhibits a hidden feeling of superiority and a poorly disguised criticism of the habits of the *new* immigrants who have dislodged them not only from their dismal apartments but also from the East Side folkways that produced theaters and shops in profusion when they were greenhorns. Not that they are heartless men or even snobbish in their approach to others more unfortunate, but that the others *are* unfortunate compared with their own desires and accomplishments they have absolutely no doubt at all. Yet they are far from that lack of understanding attributed to Jewish intellectuals by Bayard Rustin:

> I myself recently had a great number of my Jewish friends telling me "We lifted ourselves by our own boot-straps. We were intellectuals. We were this." They forgot that they lifted themselves not because Jews were well read or intellectual, but because the objective situation permitted them when they got off the boat, not even speaking English,

> but in five minutes had work because the society was buying muscle power, and therefore Negroes resent you saying, "Do as we did," in a totally new situation.

I doubt seriously that any *Jew on this earth ever told Rustin anything of the kind.* That is what he *thought* they said, but he does have a point.

The Jewish-Negro comparison is not a salutary one in the United States. Here are two peoples with two totally different approaches to living. But the "life style" of a people has very little to do with income a specific family enjoys at any given juncture in its relationship to the whole community. Two examples from my own personal experience will illustrate this point better than will a thesis in sociology. During the Depression I lived on East Twenty-Second Street in New York, in an apartment adjacent to that of a Frenchman and his wife, who had been brought to this country by a glass firm that was trying to introduce carved glass for window displays. The Frenchman was a specialist in the craft, and he had brought a whole crew from his country to help out. He was then earning $8,000 a year and his wife's income was exactly $4,000 a year. Their combined income was a munificent sum then and handsome even now. Yet, when I went into their bathroom one evening I was shocked to find the tub, removed from its mountings, full of wine that he had made himself. When I questioned him, he replied that Americans endanger their health with too much bathing and, what is more, wash away the delicate "bouquet"—that is the word he used—of their womenfolk. It made him shudder to think that the American female preferred *eau de Croton Reservoir* to *eau de cologne.* Some years later when I was in Paris with my friend Harry, his wife Diana told me that the hotel maid had commiserated with her because her young husband must be suffering so. Asked what she had in mind, the maid told her she was sure that Harry was terribly ill because he changed his shorts so often.

The fact of the matter is that overcrowding, lack of modern sanitary facilities, and the like are not necessarily conducive to bad health or delinquency, as is so often preached and touted—*as if* it were true. Sociologists usually find what they set out to discover. Today they see a one-to-one correlation between poverty, crime and disease. Writing about the crowded conditions Jewish immigrants lived in on the East Side of New York at the turn of the century, Nathan Glazer reports: "Jewish families were larger than other families, and the Jewish death rate was lower. The death rate for Jewish children under five was less than half of that in the city as a whole." And, "The proportion of Jews in penitentiaries was much below their proportion in the population." (*American Jewish Yearbook*, 1955.)

The French life style is as different from the American middle-class life style as the Italian is from the Swedish and the Jewish is from all, especially the Negro life style. Not to recognize such differences is to create a climate of suspicion and resentment that is most noticeable in the confrontation we are concerned with here. Kenneth Clark is probably correct in his opinion that "studies would bear out that most of the whites with whom he [the urban Negro] comes in contact are likely to be Jewish."

Be that as it may, Mr. Rustin may as well ask for the moon as for a $100-200 billion program as a solution for the problems that he thinks beset the Negro community. One cannot change a people's life style with money alone. One must first be sure that it is desirable to change it; then money, about a thousand years, and perhaps also a totalitarian regime will do the rest.

The Negro is usually a Protestant, but he does not share the Protestant ethic, to which Jews take as easily as a man going through an open door. Sociologists since Max Weber are agreed on this point. Achievement-oriented Jews regardless of learning or lack of it, will prosper more in an open society like our own than will almost any other minority, except perhaps the Japanese, or Chinese who are similarly oriented. Other groups will not torture themselves with denials in an effort to compensate for this peculiar ability of the Jews and their own peculiar lack of it. The Italians provide a much better minority group for comparison with the so-called Negro "peasant" migrating from the South to the northern central city.

When James Baldwin recites the litany of how *he* toted the bale, how *he* built the roads, how *he* laid the rails, how *he* dug the pits, and how *he* bent his back to build a rich America he is intellectualizing an experience he never really shared. On the other hand, an Italian intellectual, Mario Puzo (*The New York Times*, August 6, 1967), tells with good humor how *his* people—real peasants all, illiterate most of them, with nothing but their strong backs and shovels to assist them—came to the United States and did all, or most, of the same things. The truth is that both groups have contributed, as has every other minority group that is part of this nation.

But the Italian—the true peasant from Sicily and Calabria—did his work and lived in slums so crowded that it pained one to see them; some Italians still live there, never calling them "ghettos," a term derived from their own language. They are content—more or less—to be here, although fewer than 5 percent of immigrant and first-generation Italians attend institutions of higher learning, compared to more than 60 percent of Jews in similar circumstances. There are no Italians on the Supreme Court yet. But I am sure they will get there—and so are they.

From 1900 to 1910 about 2 million Italians, 99 percent of them illiterate or semiliterate peasants, arrived in this country and settled in its urban centers, especially New York. Italian family patterns did not break down, however; such a suggestion is almost ridiculous in view of Italian migration and immigrant habits in the New World. Italians stand firm, refusing to give up their neighborhoods, resenting social workers and the pull to the suburbs as attempts to subvert their way of life.

The Italian does not seem to care what his neighborhood looks like, as long as his apartment is clean, the family is together, and the table groans with food that mama has prepared with love. Only about one-tenth of 1 percent do perhaps belong to a criminal element. These Italians do not seem to resent *their* slums, *their* congestion, *their* rats, and *their* household pests. They have as many such problems as does any neighborhood in the big city but certainly fewer than civilized cities like Paris and London. Are they therefore *better* than the Negro who does complain? No. Are they worse because a few of their numbers are gangsters. No. But they are different. As Italians are not to be identified with the Mafia, neither should all Negroes be identified with the hoodlums, crackpots, and incompetents among them.

The immigrant Irishman did not succeed as easily as Rudyard Kipling suggests in the following rhyme:

> *There came to these shores a poor exile from Erin;*
> *The dew on his wet robe hung heavy and chill;*
> *Yet the steamer which brought him was scarce out of hearin'*
> *Ere't was Alderman Mike inthrojucin' a bill.*

All this argument is intended only to establish that different minority groups blend into the national scene in different ways.

to make a patchwork; but the *melting pot,* as Glazer and Moynihan have shown, has never yet become hot enough to homogenize the varying ethnic and national identities that make up this country. The Italian and Irish experiences—and they can be supplemented with comparisons from Polish, Hungarian, and other groups of peasant origin—are significant in that they demonstrate that *peasants newly arrived in the big city do not show the kind of family disintegration that has been claimed.* In fact, the family is the one cultural pattern that is hardest to break down. The Jews, who have wandered the face of the world; who have been persecuted as no other people has been; who have been subjected to a thousand cultural atrocities; who have been converted by force or chosen death; who have often abandoned their religion in their quest for modernity; who have produced some of the leading sociologists in the world; who have been tempted, tried, and tested in a thousand crucibles have survived more *because* of their close family relationships than because of any other factor, including their religion. The family is the cement that binds the Jews together. A similar (not identical) cement binds the Italians, the Chinese, the Poles, the Japanese, the Hungarians, and the Swedes in the United States.

We are back to the Negro family; rather we are back to the Negro family that lacks a father at home. It is significant that this kind of Negro family is different from any other family that this country has ever known, and no amount of socioeconomic explanation will alter that fact.

There is "collusion" of silence on this point among the ignorant and those who pretend that they do not know the truth because they fear a bad reaction from their friends, a bad press, or both. At least the Negro *knows* the truth, even if sometimes he acts as if he does not. Sometimes in collusion with the white liberal—and how the Negro intellectual despises him for joining this collusion—he enters the world of pure fantasy and *wishes* that it were not so. Then all would be right, and a few extra dollars would be enough. When E. W. Burgess argues that the Negro family is the way it is because slavery made stable family relationships impossible and the pattern established under slavery still persists—at the same time he argues that there is a "qualitative" difference between white and Negro families—he is guilty of a contradiction.

When this concept is repeated *sotto voce* by white enthusiasts it can be regarded as a form of patronization the Negro can do without. The white reporter who, in an effort not to offend the important Negro leader whom he is interviewing, dismisses the riots as the "unimportant" acts of a "small minority" commits the worst kind of offense to the Negro by failing to point out that the "sanctioning of larceny by racism" (as Professor Abba Lerner says) is the most injurious kind of paternalism that the white man can practice.

Would it not have been better for the reporter to reply to the Negro leader, who had mentioned the "small minority, not more than three percent [about 600,000] were involved," that even that much hoodlumism or rioting or rebellion is too much for the nation to overcome? He might have added: "As long as you minimize the enormity of the transgressions and in fact seek excuses when punishment is required, you are betraying the other 97 percent as surely as if you had told them they were all hoodlums."

Those who argue that the Negro family should be fortified with extra funds, so that the father-husband will have an incentive to stay home, misconstrue the nature of the Negro family as it exists and as it has existed for some time past. Lee Rainwater, in an interesting study, "Crucible of Identity," tells us that "in the city the woman can earn wages as well as the man can and she can receive welfare payments *more easily then he can*" (emphasis mine). Rainwater is convinced, furthermore, that "lower-class Negroes know that their particular forms [of family life] are different from those of the rest of the country. . . ." It is very likely—as I think Ralph Ellison was trying to tell Daniel P. Moynihan—that they find it natural to reinforce *their* conception of the good family, whatever anyone may think about it, by going on relief. Relief does not hold for them the kind of stigma that it holds for most American families brought up with different values, in which receiving charity and public aid seems parasitic. If the Negro does indeed regard his "female-headed" family as just as good as that of the white man, then no amount of money in the world will induce him to change it. Actually it turns out to be a problem only for white gentlemen with so-called "middle-class values" tinged with egalitarian beliefs and the gnawing desire to right all wrongs. Before righting wrongs, one ought to be sure it *is* a wrong he is righting.

It is certain that, as long as the Negro prefers that kind of family—and it seems that the so-called "matriarchal" or "grand-matriarchal" family is not confined to lower class Negroes at all but is widespread among all classes of Negro society in the United States—who are we to tell him he should not?

Dr. Alvin F. Poussaint, a Negro psychiatrist, says that Negroes riot to assert their masculinity because they resent having been "castrated" by the white man. If that is true, then the riots will continue, for a "castration complex" is not disposed of in a day. The main explanations of riots are contradictory. According to one kind of explanation, rioters represent a small fraction—less than 1 percent and 3 percent the lowest and highest estimates—of the Negro population, and therefore the problem should not be magnified out of proportions. Psychiatric interpretations like that of Dr. Poussaint seem to imply deep-seated resentment among large numbers of Negroes. Robert Conot's *Rivers of Blood, Years of Darkness,* a study of the Watts experience, which was at first described as the work of a small group of malcontents, suggests that there really were large numbers of participants. In a review of Mr. Conot's study, social critic Richard M. Elman speaks of the "large numbers of the deprived [meaning Negroes] . . . willing to risk the little they have, for the momentary pleasure of keeping white authority at bay."

Dr. Poussaint, however, goes even further and suggests that the sense of being "castrated," among all the Negro male's frustrations, is the one that accounts for his failure to head up his family. Even if this sense were a result of slavery, it would still not be possible to accept slavery as the entire explanation. It seems doubtful that the Negro who has been able to overcome almost all the traumata induced by slavery should cling to this one pattern.

For instance, we know from Carter G. Woodson, who has delved with a good deal of sound scholarship into the subject of the "free" Negro family, that as early as 1830 among thirty-two family groups he investigated in Richmond County, Georgia, most of them well-to-do by any standards, *twenty of the thirty-two families* were headed by women! Yet we learn that in an adjacent area, where property relations were stronger, that is, where wealth was counted as property possessed, the family situation was considerably more stable.

When E. Franklin Frazier speaks of the "free and uncontrolled" behavior of the "larger Negro world," it may be that he is torn between opposing values; he speaks of the "free" behavior of the majority and deplores the "exaggerated" valuation put on moral conduct by certain puritanical upper-class Negroes, yet he also deplores the "waste of human life, the immorality, delinquency, desertions, and broken homes which have been involved with development of the Negro family in the United States."

Dr. Frazier did know—and refused to hide from his readers—the one truth that is repeated over and over in studies of the Negro and his relations to the world around him: "*In general, home-ownership . . . offers the best index to the extent and growth of the stable Negro family.*" I italicize this statement because it might in the long run be less expensive to provide the Negro with real property than to provide professionals with unlimited sums to spend from the national treasury, that is, from the Negro and white working, middle, and upper classes of both races who are almost all real-property owners.

However corny it may sound, a man who has a home of his own is generally a much more law-abiding citizen than a man who has no property.

We have seen that during the riots in our cities even the black hoodlum has often been respectful of the property of a "soul brother."

Very few Negroes are racists, fewer even than whites. But, as has so often happened in the United States, the Negro seems doomed to suffer the most imcompetent, self-seeking, and mendacious leadership that any people in the entire world has had to cope with. In passing it is safe to say that all the leaders of the so-called "civil-rights" movements, including all the black-power boys, do not represent more than 5 percent of the Negro population in this country, and I doubt even that figure. The best Negroes, with a few exceptions, simply do not go in for that kind of thing. Why? Perhaps the reason lies in the fact that most Negroes do not think that they have the *special* problems conjured up for them by interfering white liberals—with Jewish liberals in the forefront—and greedy black braggarts who, often enough, lead no one but themselves. These leaders have learned through the years how to make do without holding jobs that require any more competence than collecting other people's money and preying upon imaginary ills or evils that only they—in their leagues or societies or committees—can help to solve. Negroes usually know them for what they are and hardly contribute to *these* efforts; instead rich white men (especially Jews) are corraled into these movements and resign only after the most humiliating insults make it unbearable to remain.

I recall how the famous folk singer-actor Theodore Bikel resigned with a huff from S.N.C.C. only after that organization had come out for the Arabs in the Israeli-Arab war. He did *not* resign when one of the group's leaders threatened to shoot any "honky"—or the President's wife—on sight, incited to riot, and called for the burning down of Cambridge, Maryland. Mr. Bikel resigned only when *his* people came under attack. Just who does he think *his* people are? (One distinguished gentleman and a former member of the Israeli parliament told me to say in this connection that he, for sure, is not one of Theodore Bikel's people.) I also recall with distress what Pete Hamill, the New York *Post* columnist, once reported about the marches of Mississippi: "I heard several people sneering at the 'poor Jews coming down to whip themselves' by working for SNCC." With Mr. Bikel, Harry Golden, author of *For Two Cents Plain* and Rabbi Arthur J. Lelyveld also left S.N.C.C. What were these gentlemen doing there in the first place? Getting themselves whipped?

Meanwhile, it is worth observing that, as far as Negro stability, education, and many values that other Americans cherish—even if only as ideals—are concerned, improvements will not come until we all understand, and the Negro makes peace with, the problem of the Negro family.

Are there any solutions? I do not really know. But the following suggestions may be of help to the *white man* in the United States; and because this area is volatile we shall enter tentatively, leaving as soon as asked. With a good deal of humility but with little deference, I suggest that THE WHITE MAN DO NOTHING ABOUT THE PROBLEM!

Surely it is foolish to intrude values, however highly cherished, where they will be rejected or misunderstood by those who either do not want "our" help or are convinced that only dirty back yards, rats, slum tenements, and poverty encourage fatherless families among urban Negroes. Others are equally convinced that lack of "family discipline"—as they understand it—is what causes blight in the so-called "ghetto." Because I think these explanations are incorrect it is my intention to examine this whole business of poverty, delinquency, illegitimacy, goals, and achievement. But before we turn from specific discussion of the Negro family let us note that just about the only valid parallel is not to refer to imaginary and nonexistent similarities between Danish, Scottish, and Irish families as both Moynihan and Rustin do. There is, however, one comparison and that not a too rigid one between the Jewish family and the Negro female-headed family that has so far seemed to escape the paralogistic passions of the socio-economic and deterministic scholars and special pleaders.

I refer to the Jewish family *left behind* somewhere in East Europe for as long as five or ten years, sometimes even longer, while the male head of the family went off to America to seek his fortune. Left to her own devices in the old country until her husband could raise the *schifskart gelt* (passage money), the mother raised her brood of three, four, or five children, a matriarch in total control of her demesne and saw to it that—in effect—her *fatherless* children did not become dropouts, delinquents, hoodlums, or neurotics requiring special programs to catch up with more favored children from complete homes.

Many a Jewish boy and girl in the early years of the century, when East European migration was enormous, first saw their father, in any meaningful sense, when they were quite grown up—sometimes fifteen years old or older. Yet these children did without fathers in their *formative years,* as the social workers would say. They had only mothers or grandmothers to guide them. They ultimately rejoined fathers who were real strangers to them, in a land where they did not know the language; they went to school, studied, played truant sometimes (but not too often), got into scrapes (and cried too often), and became union men, gangsters, playwrights, bosses, intellectuals, businessmen, athletes (not many), professors (too many), entertainers, showmen, publishers, radicals, patriots, traitors, soldiers, and scholars.

I am not trying to make out a case for the Jewish family, whether headed by mother (in Europe) or by both parents in this country. In a very profound sense, there is nothing so abrasive as the Jewish father-mother relationship, the bickering, badgering, blaming, and *broygez*—that special Jewish quality of having a mad on that can last for years and in which the mother will not talk to the father (or vice versa) for a lifetime. An East European Jewish mother's telling one of her offspring to "tell papa to eat his soup before it gets cold" evokes yaks of recognition so predictable that Sam Levenson has built a fortune as a comedian on them. One Jewish intellectual (Milton Himmelfarb) has even tried to explain the flight of the Jews before the Negroes in terms of *broygez.* The Jews, he argues, do not fight back as do other minority groups like the Italians, who usually stand their ground in their neighborhoods, or the Slavs in the Midwest (Milwaukee), for example, who meet Negro encroachments with baseball bats and worse. The Jew employs his *broygez* to sulk (God forbid!) not, to fight back. Mr. Himmelfarb asks plaintively why it is not better to be mad at one's neighbor than to hit him? He thinks it is better. He does not realize that the Negro would have less contempt for him if he would simply shut up and fight back. "Didn't we send Schwerner and Goodman to Mississippi?" he asks. The Negro replies, "We couldn't care less." Horace R. Cayton, co-author with St. Clair Drake of *Black Metropolis,* a book that is much quoted on this topic, once said, "There is . . . resentment on the part of the Negro that the Jew is so often willing to fight the Negro's battles, but is often reluctant to fight his own."

The kinds of battle the Negroes are fighting, the things they want—or the things it is said that they want—the solutions proffered, and the strategies outlined for them by radicals of CORE and SCLC differ in no essentials from those suggested by more moderate leaders like Roy Wilkins, Whitney Young, Jr. (until recently), Bayard Rustin, and A. Philip Randolph; nor do they differ from those of the exponents of black racism and violence. At least in the latter instances, we can recognize the danger to the entire Negro community in the overt expression of, to put it politely, impatient young men. We can recognize it and act accordingly.

In Dr. King's demands for massive "dislocation" of the economy and education of the nation to bring more jobs and better schooling, we faced the much more dangerous problem of a man who behaved as if the Nobel Peace Prize were a grant of immunity that permitted him to advocate an anarchy infinitely more conducive to race hatred than a hundred riots in the streets would be. The important problem is to determine just what Martin Luther King wanted—and what his more moderate colleaques want when they call for action *now.* They want more funds *now;* they want more legislation *now;* they want more demonstrations *now.* For what purpose? Why? What will $100 billion dollars do that half that amount has failed to accomplish? What will another civil-rights law do that all the others have failed to accomplish? It is possible that Negro psychologist Kenneth Clark is right when he says, "Riots increase in frequency with an increase in civil rights legislation." Why? "When government promises people things which are not delivered, you are increasing their rage. . . ." I hope that he means that *government* has no business promising anything that it cannot deliver and that no government in a free society should be able to control the lives of people to the extent that it can only *increase their rage*. (Dr. Clark calls on the business community to help give Negroes jobs as "an investment in stability.")

The more money that is poured into the so-called "ghettos," the more delinquency and the need for relief will be the lot of the fatherless family. And less likely, too, will be reinforcement of the family, without which all other panaceas for "lifting the Negro up" will surely fail. For instance, when Mayor John V. Lindsay proposed to alleviate the plight of families on relief in New York by offering a $3.5 million program "to permit welfare mothers to work full-time while their children are cared for by other mothers on relief," he was as likely to splinter further the fragile relationships within the Negro "family" as if he had set out to bust them up. The mother is *the* binding force, responsible for whatever decent and kindly behavior is exhibited by her offspring. If the mother is removed from the fatherless home, even that one source of comfort will be stripped from the Negro child. The Mayor is an impressionable man; he is impressed by the philosophy of the buck and what it can accomplish in an affluent and largely unthinking United States.

When Lyndon B. Johnson dropped the ball passed to him by Daniel Moynihan, Senator Robert Kennedy picked it up and started an end run for a touchdown. He was the only man in public life at that time who saw the problem of the family as basic, underlying all aspects of the so-called "Negro" problem.

Like most men who have given the subject more than passing attention, he recognized the moral erosion inherent in continuing relief on a massive scale, as an end in itself. Some families in the nation's Harlems have been on relief for two or more generations. Fatherless families have increased in number, children have multiplied, and relief costs have mounted.

It is a vicious cycle, Senator Kennedy argued. He wanted to bring men from the private sector of the economy into the areas affected and let them work out plans for jobs on the spot, so that the Negro's family structure—the cause of much of the delinquency characterizing Negro urban life—could be reconstructed to provide well-paying jobs for the missing fathers and thus enabling families to escape relief. At least he spoke courageously on the degrading aspects of relief for those who receive it and for those who dispense it. He was heard with some caution because he failed to spell out how it would be possible to put a man to work in local industry or business at a wage that could adequately compete with the tax-free cash a family on relief already receives, in New York, for example. There a mother with four children can receive more cash (for rent, food, household maintenance and various special allowances) than almost any man of hers, *regardless of his skills*, could match. Unless Senator Kennedy found ways to ensure such vast profits and tax exemptions to local job suppliers that they would be able to pay skilled, semiskilled, or unskilled black workers 50–100 percent more in wages than the average in the metropolitan area, his plan was almost doomed to failure.

As long as his thinking was rooted in this most fundamental of all relations confronting the American Negro community, the Senator was on the right track. But once Humpty-Dumpty is broken it is hard to put him together again. It is and will remain difficult to *remove* mothers and children from the relief rolls. The trend, in fact, is toward more relief, more illegitimacy, and more delinquency—ultimately more riots and rebellions. All the commissions now being established and yet to be established will be able to do little but make expensive reports on where we went wrong, why we failed, and the like. They will certainly fail because they approach the problem with dollar signs in their eyes instead of the compassion that requires the surgeon's objective yet essentially humane approach. If surgery is called for, let us cut with all the skill that social surgery can muster. If instead sedation is called for, let us administer it, although such a diagnosis may indicate that the disease is terminal.

I do not believe that the Negro family situation is terminal, although it is being eroded more and more each year by increasing permissiveness and greed among local politicians who think that relief earns them gratitude at the polls. Actually it brings them only contempt—and justly so.

No sooner had Senator Kennedy proposed his plan for massive business deployment in the so-called "ghetto areas" of the nation than he was attacked by Dr. Thomas W. Matthew, a Negro and

President of the National Economic Growth and Reconstruction Organization (N.E.G.R.O.). It was Dr. Matthew's opinion that the Kennedy plan was "another game in which everybody loses"; and that it represented "a double fraud, a booby trap for white industry as well as for the black ghetto."

Dr. Matthew's criticism should not be dismissed lightly. He has provided a telling demonstration of what can be accomplished by Negroes in local communities. He has taken the leadership in developing such Negro enterprises as a hospital (Interfaith) in Jamaica, New York, retail shops in Bedford-Stuyvesant and co-operative bus lines in Jamaica as well as in Watts. This has been achieved without a dime of Federal, state or municipal aid and despite considerable bureaucratic resistance. When he started the bus line in New York, he was told that it would be necessary to obtain a franchise. This could have delayed the project for years, if not for eternity. Sidestepping the obstacle, Dr. Matthew bought some old busses and sold stock in the line for 25¢ a share. Soon, thousands of shareholders had sorely needed transportation that neither public nor private transit lines had ever provided. (In 1969, largely because "established" patterns for solving "ghetto" problems had been violated, he would be harassed for income-tax violations and given a six-month jail sentence. This sentence was later set aside by President Nixon, in his first official pardon since he took office.)

But, in a plea to business and industry to come to the aid of the "ghetto," Dr. Kenneth Clark said, "Business is the least segregated, least discriminatory, most fair, of the areas in our society—better than education, religion, unions, or government."

Welfare is a trap. As long as a family of six or eight can receive more from public agencies than even the most skilled worker can earn, that family is doomed to be permanently on relief and permanently without a father at home. The nation is not as wasteful of other natural resources, like forests and wildlife, as it is of the 10 percent of its population that is Negro. The Nixon administration's proposals for a work-incentive plan to families on relief may help. How much such "help" will actually amount to in keeping the father in the home it is difficult to predict. It is however a stab in the right direction. But a hard look at the concept of "reverse" benefits—not a negative income tax—for families on relief, offering more for less, *more aid for fewer children*, may be of considerable help in reducing delinquency and the proliferation of fatherless children in "ghetto" homes. What is implied is that a welfare family with two children would receive relatively higher benefits—so long as the family was not increased—than it would if it did multiply, permitting only such additional funds to see to it that the new-born infant did not starve or freeze to death. In other words, one would make it "costly" for the husbandless mother to indulge her passions since she would not be generously rewarded with more aid for each dependent child.

Chapter Twelve
POVERTY IS NO CRIME

Richard M. Elman sums up his opinions of Robert Conot's *Rivers of Blood, Years of Darkness,* a study of the Watts riots: "When the ghettoes riot, they only promote the recognition that there are increasingly fewer avenues for cooperation between the poor and the rest of America." Now this equation of the "poor" with Negroes and poverty with crime (rioting) is a root principle of much, if not most, thinking on the subject by professional social workers, sociologists, and spokesmen for civil rights in general. Why the Negro poor should behave differently from and worse than all other members of minority groups equally as poor or poorer is a question dismissed as irrelevant. The person who asks it is usually granted an accusing stare, as if he were questioning the rights of Negroes to life, liberty, and the pursuit of happiness. Today it has become accepted practice—especially by the mass media—to equate poor with Negro, and poverty with crime.

All the world's literature teaches that poverty is no crime or shame. "The poor always ye have with you," say the Gospels, and phrases like "poor but honest" grace the folk wisdom of the ages. From Deuteronomy we learn that "The poor shall never cease out of the land." But we can appreciate the wisdom of Othello, who said, "The poor are only those that have no patience." Patience! That seems to be what is troubling so many well-meaning disciples of civil rights, who cry that, "after four hundred years" or "one hundred years after slavery," it is time that they have "theirs"—*now.* The immediate is emphasized, diminishing the achievements of a people that has, in a very real sense, started from scratch to catch up with the white world around it.

Some members have caught up faster than have others; some have *surpassed* many members of the white world; many—too many—have lagged behind. For whatever reason, it is a fact that Negroes do not do as well economically as do some other minority groups in the population. But that they do *as well as*—or better than—some others is often forgotten or not mentioned. Not all groups in this country make it equally quickly up the ladder of social and economic success. Some reach the top—or near it—more easily than do others. Some falter and fall back. Others never even try to "make it," refusing to make the sacrifices required for certain kinds of achievement.

It is well known that the Jewish minority, representing only 3 to 4 percent of the total population, has achieved more in goods, services, and other emoluments than almost any other single group in the land, even including the dominant white Protestants. It is by now an open secret that in 1957 the American Jewish Committee interceded with the Bureau of the Census in Washington and besought it not to ask questions about income related to national grouping in the 1960 census, for fear that the comparatively high income levels of the Jewish minority would lead to anti-Semitic outrages. The Bureau complied. Perhaps the Jewish organization was right in its demand, for the bare figures could not reveal the amounts of sacrifice that it took for many Jewish families to send their children through high school, college, and graduate school. Jewish parents have often been ready to sacrifice their own immediate needs and luxuries so that their children can pursue education; such sacrifices often bring a kind of happiness that members of other minorities who are not so ambitious, not so avid for success, have not always understood. Jewish boys and girls attend institutions of higher learning at a rate ten to fifteen times above the average for sons and daughters of other immigrants; one reason is that their parents' life-style puts a tremendous premium on learning for its own sake.

That education also helps the Jew to climb the ladder of success in the United States where learning is rewarded to a degree unknown elsewhere in the world—sometimes makes him the butt of his neighbors' ill humor. But comparison only with Jews is unfair to Negroes. No other minority—Italian, Polish, Swedish, or Irish—equals the Jews in this area. Yet, when statistics are cited, almost invariably the Jewish achievement is the one juxtaposed against the Negro achievement, and the gap is indeed enormous. No one, at least to my knowledge, has ever compared the levels of educational achievement of the Polish, Hungarian, or other East European minorities with that of the Negro minority. But it seems likely that such a study would show that Negroes have not done as badly as has often been suggested.

It would also be false, however, to leave the impression that the Negro should be satisfied with his achievements and should shut up. Neither should other groups who find the upward climb to success a bit fatiguing. All these minority groups are not prepared to make the kinds of sacrifices that the Jews have made in order to succeed. Whether or not they have the capability is not the point at issue. Not every man is happy pursuing degrees and the professions. In fact, most of the population has a rather low regard for the intellectual who seems never to be satisfied with the degrees that he can earn in a lifetime of self-denial.

A degree and a modest bank account hold little attraction for those of the majority who believe that a night at the fights or the ball game, or a weekend fishing, or playing cards with cronies is as satisfying as pontificating on no-audience educational television. The average man watches "Bonanza," a wrestling match, or a ball game with a bottle of beer in one hand and an overstuffed sandwich in the other. No Negro worth a dime would give up his Saturday night—the Negro Saturday night that only a William Faulkner or a Claude Brown could describe with any degree of adequacy—for a bull session on the proletariat or the latest trend in the novel of the absurd. (I recall the pre-Christmas moralizing of an editorializer on network television, as he deplored the sad holiday that awaited the down-trodden Negro who would not be able to enjoy the kind of Christmas *he* was going to have. I was sitting with a Negro friend and his wife, and they both burst out laughing at the absurdity of his commiserating with them; they knew that the poorest Negro in Harlem would enjoy festivities on that Christmas Day that would make the editorializer's celebration seem mean by comparison.)

It is correct that the white man does not understand the Negro. The white man surely does not "dig" his way of life. William Faulkner, who was never overweening in his approach to the Negro as a living person or character in a novel, understood him in a way that few, if any, white intellectuals do. But the black man knows how to con the white man, especially when he thinks that the white man is showing him too much attention, as Jewish intellectuals usually do. When he is handed an ideological line for free, called an "exploited" member of the working class, and told that he should join forces with the white working class, he often remembers that the white working man has been the major obstacle to his achievement, in unions and on the job. He recalls that in the South he has usually been most despised by the "poor white trash," that southern workers organized groups known as Red Shirts (after Garibaldi) to smite the blacks (or scabs), and that in the South the Civil War was called the "poor man's war" by the working men. He recalls, or soon learns, that populist demagogues counseled the white working man to keep the "nigger" in his place and that even the humanitarian socialist and friend of the working man, Eugene Victor Debs, wept when he heard of the demise of professional "nigger hater" Tom Watson of Georgia. He hears that Jack London called himself "a white man first then a socialist." Only thirty years ago the Communists told him that he should live in a separate nation in the "Black Belt." His anger is rekindled when this same separatism is promoted today by men who speak a jargon that troubles him, as it troubles the nation.

Man for man the Negro worker earns about three-quarters what the white worker earns in the United States. The Negro wage earner is thus less well off by 25 percent than is the white wage earner. He is still the second-highest-paid man in the world, in terms of real earnings. His children go to college in *greater numbers* than do the children of Englishmen, Frenchmen, Swedes, Finns, and Norwegians. Between 3.5 and 5 percent of boys and girls go on to universities in most European countries. In America this figure has climbed to a whopping 40 percent of the youthful population. Jewish boys and girls register even higher percentages. But Negro boys and girls probably do no worse in this area than do the boys and girls of other immigrant groups. The Negro is thus not so poor as to be pitied, not so indolent as to be pilloried, nor so different as to require the laws, the commiseration, and the contumely directed at him by his benefactors and detractors. Much of the Negro's difficulty lies in what the white liberal expects from him. The banal term "revolution of rising expectations" is often cited on his behalf. But the Negro's expectations cannot be fulfilled until he *wills* and strives for fulfillment. In the Spring, 1967, issue of *Freedomways,* the self-styled "organ of the Black Liberation Movement," a young writer from Tuskegee Institute tells the white man to stop being solicitous about the black man's cultural deprivation. He argues that he does not mind being deprived of the white man's culture as long as he has his own to admire.

At predominantly Negro Fisk University, student leader Wilbur Hicks is quoted as saying: "Carl Sandburg, Bach and Beethoven are closed subjects among students now. We recognize that perhaps you ought to study them, but we don't really relate to them. There's no blackness there." He is probably right. Yet what a sad confession. It is a pity that *blacklessness* has become so suspect among the younger Negro generation that it is willing to throw overboard Sandburg and Beethoven and Bach. But that is what many of them want. They should be allowed their new black dreams—but not at the expense of white nightmares. And the new appeals for black quotas in colleges based on "open enrollment" will, indeed, bring more young Negroes into our universities—and prepare them less for life in our society than the previous "exclusion" based on merit. Before, those who were competent made it—or had a chance to make it. Now, even this will be denied them because the Negro student will never quite know if he is in college because of his color or his ability—and this could hurt with a deeper pain than that previously experienced.

There are enough Negro millionaires around, however, to make even the least envious whites open their eyes. There are enough Negroes who have "made it," despite talk of castration, violence, and slavery, to challenge the impudence with which the good white man usually approaches Negro problems. But there are too many Negroes who have not made it at all for whatever reasons. Certainly it is not news as Nathan Glazer and Daniel P. Moynihan point out in *Beyond the Melting Pot*, that when the Negro does go to college he does not prepare himself for professions he thinks are closed to him (as Jews and Japanese did). (The same, by the way, is also true in Brazil, a nation never accused of harboring a racial bias. Yet here, too, we find that only about 1 percent of the Brazilian blacks make up the quota of professionals—doctors, lawyers and engineers.)

To point out that in the last fifty years the number of Negro physicians has not increased commensurate with the opportunities provided is to miss the point, for it is still doubtful that Negro doctors in large numbers would attract enough white patients to compensate for their efforts. That is, of course, not the reason why Negroes do not enter engineering or mathematics. Perhaps they do not want to; perhaps they cannot. Whatever the reason, it is a fact that, without tradition of family discipline and community expectations, Negroes will produce no more professionals than they have been producing for the last 100 years, and their frustration will surely be intensified. We shall go into this point when we take up the whole matter of Negro education. But it is relevant here that Booker T. Washington's prophecy for his race almost three quarters of a century ago is still valid today.

He wanted the Negro to acquire skills in the mechanical arts and to shun degree hunting as an idle endeavor of those who had no need to work hard for their share of the nation's goods and services. Opportunity, according to him—the wisest of Negro leaders—was there. It had only to be grasped. He knew that man is essentially *unequal* in talents, abilities, and ambitions; each must have a chance to satisfy his own needs in his own particular way. But a *chance* he demanded, with no ifs, ands, or buts. Give Negroes a chance to further their own interests was his essential appeal.

If white society had not persisted in intruding its values into this area much mutual suspicion might have been avoided. Worse, some well-meaning people have insisted on creating a double standard of behavior, offering the Negroes what their best spokesmen have always repudiated. Mr. Washington could say, "I never see a dirty back yard but I want to clean it up." The Negro today is told that it is the duty of the slumlord, the city, the state, the Federal government, or all together, to clean up the back yard. (It is not uncommon to find that local governments are spending special emergency funds for such purposes.) Finding debris is a sort of a racket, the yards soon fill up with more garbage than the can provided can hold, and a new drive—at an hourly wage—is started to "keep our neighborhood clean."

These projects are usually popular in the summer because the mayors of our cities, jittery over their political futures, fear riots. As plenty of money—local, state, and Federal—was generated by past riots, some people expect the amount to rise with each new riot, or threat of riot. Blackmail breeds blackmail, and, when decent, law-abiding citizens read that local troublemakers, rioters, and five-percenters are paid $50 a head—$50 *per diem* was paid to some of the most unruly elements in the Harlem area in 1967–68 for serving as "advisors" to the Federally sponsored antipoverty program!—they are so indignant that, as one man told me, "The next time the riots start here, I am joining in, then maybe they'll pay me off with fifty dollars a day." Why not?

The notion that every poor man—especially every poor Negro—is an incipient crook probably originated with the very rich who have held office in the land. It is perhaps impossible for a Nelson Rockefeller or a Ted Kennedy to imagine how it feels to be poor. There may be a deep-seated (but well-suppressed) feeling among those with inherited fortunes who go into American politics that their parents or grandparents made their money without too many scruples and that others probably do the same. After all, they have been educated in universities where such learning was considered the latest fashion in sociology.

They may have learned from Professor Max Scheler that "moral indignation" is peculiar to the middle class psychology and represents "a disguised form of repressed envy," as Daniel Bell says, invoking Dr. Scheler to support his thesis. Crime is therefore not a matter of right or wrong but merely a matter of power and expediency. It is best (safest) for the middle class to be moral; therefore it is moral. If it served the middle-class to be immoral, the middle class would be immoral. Right and wrong, moral and immoral, good and bad are only the self-serving devices of a society that does what it has to; all that poppycock about morality is a lot of bourgeois nonsense.

Not that Scheler or Bell oppose safety in the streets, at least in their streets, or in the home, at least in their homes. But they are, of course, repressed middle-class intellectuals. Were they as brave, as courageous, as uninhibited as those who do not enjoy middle-class respectability they too might be out smashing windows and grabbing color-television sets.

The moderate Dr. Bell may yet find himself the hero of the Stokely Carmichaels and the H. Rap Browns if they ever explore the implications of the *End of Ideology*. Maybe Eric Hoffer, the longshoreman philosopher is right when he says, "A society that can afford freedom can also manage without a kept intelligentsia." Certainly when one reads pornographic attacks on the Puerto Rican community by an American cultural anthropologist; when one ponders the advice of a Columbia University professor of social work that the people of Harlem should so overload the relief rolls that the entire fabric of the city will collapse; when ministers and rabbis hire a professional revolutionary at $25,000 a year—a man whose heroes are Che Guevara, Malcolm X, and Stokely Carmichael—to "upset," "enrage," and "disorganize" peaceable communities, then indeed a *free* society is in danger from its paid intelligentsia, from subverters of the social order on its Federal and university payrolls.

It is the "bloody-minded professors," as Winston Churchill called them, who preach equality above law and both above freedom. Alexis de Tocqueville warned that a society that values equality above freedom is sure to lose both. Hannah Arendt has said, and she *is* speaking of the Negro: "No doubt, wherever public life and its law of equality are completely victorious, wherever a civilization succeeds in eliminating or reducing to a minimum the dark background of differences, it will end in complete petrifaction and be punished, so to speak, for having forgotten that man is only the master, not the creator of the world."

"Poverty" is a political term. People have been *poor* throughout the ages without having been aware that they lived in *poverty*. Former President Dwight D. Eisenhower once remarked that he had not known that he was poor until he had earned enough of fame and fortune from the world to be aware of it. Sam Levenson, himself a product of East Harlem when that area was largely populated by Jewish immigrant families, tells us that only in later life did he learn that he had lived in a "depressed area." Maybe the area was depressed, but he was not, and neither was his family. In a more serious vein, Irving Howe, the eminent social critic, recalls his own poor boyhood in the Bronx: "To be poor is something that happens; to experience poverty is to gain an *idea* as to what is happening" (emphasis mine). The italicized word implies an awareness gained via ideology. And ideology, any ideology, can become an evil that feeds on itself, producing ever greater evils as it spreads.

Are the Negroes in America poor? Compared to any other people on earth, excepting *some* of their white counterparts, no. Are they well off? In terms of the *average* earnings of the white wage earner, against the answer is no. But those industries where they do the same skilled work as do white men—in steel, in autos, in transportation—the answer is yes. The problem is that not enough Negroes are skilled and that not enough unskilled Negroes are employed because a benevolent officialdom has legislated the minimum wage. Although a minimum wage seemed desirable a half-century ago, today it results in low employment among the unskilled, a burden that falls most heavily on the Negro population.

A survey reported in *The New York Times* of Feburary 13, 1967, under the heading "In Mississippi Delta, More Pay Means Fewer Jobs," showed conclusively that, because of the new Federal minimum-wage law, thousands of agricultural workers in the South were losing their jobs. Ralph K. Winter, Jr., Associate Professor of Law at Yale University, commented in *The Times* on July 27, 1967:

> That minimum wage laws must raise wages effectively only by creating unemployment seems quite apparent and has been recognized by such politically diverse economists as Prof. Milton Friedman of the University of Chicago and Prof. James Tobin of Yale . . . That such unemployment will fall on the groups which have fewest skills and are most discriminated against, the ones, that, who can least afford it, also seems clear.

A "flashback" is in order. In 1962 a National Conference of Small Business was held in Washington, D.C., to deal with small business and the Negro community. A major report at the conference pointed out that "business careers do not rank nearly as high as status symbols as other professions." It also noted that Negro colleges almost never offer courses in business administration. Representative Charles C. Diggs of Michigan, general chairman of the conference, expressed the view that the "central concern" of the meeting was "greater participation of Negroes in American business life." What was not emphasized by this well-meaning legislator—only a half-dozen years before Negroes burned and looted *other* people's businesses—was that the Negro small businessman's opportunities had been largely curtailed, thanks to a law he himself was then sponsoring. Mr. Diggs was working to impose on the nation a law that requires a Negro who wants to start a small business with a couple of employees to pay them a minimum wage out of proportion to the small capital that he is likely to have at his disposal. Larger retail and small manufacturing firms were hard hit by this law, which largely serves the interests of the big unions and big business, to the disadvantage of unskilled workers and small businessmen. The conference, by the way, was the first and so far the last of its kind held in the nation's capital.

Since local politicians tend to look upon Washington as a cow giving endless milk, the Negro who complains of too many promises broken can be told that Washington failed to come across or did not give enough.

This word "enough" is troublesome, for it means different things to different people. The reason I bring up this obvious point is that some people apparently think "enough" can be measured on a yardstick—or even a micrometer. Daniel Moynihan was praised for his remark, "We seem unable to recognize that to be poor is not to have enough money," but I do not understand what he means. In Mr. Moynihan's terms we are all poor. How much is "enough"? Who, indeed, has *enough* money? A Maharajah in India, H. L. Hunt, Nelson Rockefeller? What is enough *money?* Implicit in the query is the question enough for *whom?* Some Negroes have more money than the average white man, but the majority have not. The problem troubles white liberals and supporters of yesterday's struggle for civil rights and integration. Where do we go from here? Perhaps it would be well to review some of the Negro's ups and downs since Emancipation.

Some of the patterns visible now, in what may be described as the "New Reconstruction era" of the Negro people, were observed by Booker T. Washington a long time ago: "During the whole Reconstruction period, our people throughout the South looked to the Federal Government for everything, very much as a child looks to its mother." The only change is that today almost *everyone* looks to Washington as his collective mother. Booker T. Washington knew that such dependence "so far as it related to my race, was artificial and forced." What disturbed him then—as today it disturbs whites and Negroes who are not prepared to scuttle the nation in favor of a totalitarian ideology—was that "The general political agitation [under Reconstruction] drew the attention of our people away from the more fundamental matters of perfecting themselves in the industries at their doors and securing property." The "fundamental" problem is still much the same.

Commenting that, when freedom came, "the slaves were almost as well fitted to begin life anew as the master, except in matters of book-learning and ownership of property," Washington notes that "the slaves, in many cases, had mastered some handicraft, and none were ashamed, and few unwilling to labour." Unfortunately, Booker T. Washington has been held in low esteem—especially among liberals who are disappointed that this ex-slave showed so little bitterness at his former condition. Yet the character of the man shines like a beacon, and his image is more appropriate for today's young Negro to emulate than those of imaginary heroes of the African past, which is as irrelevant to him as it is to any other American.

In the post-Reconstruction crisis the Negro had a choice between the simple wisdom of former slave Booker T. Washington and that of Harvard-trained educator W. E. B. Du Bois. Dr. Du Bois died a Communist, a proponent of the totalitarian philosophy of Kwame Nkrumah of Ghana.

In a sense Henrik Ibsen was right when he observed that "Liberalism is the easiest philosophy for a man without character." Character is what stands out in all its tough ruggedness and gentleness in Booker T. Washington's autobiography. Who is ready for freedom? Only the very young and the truly hopeful. For the aged freedom is often a nuisance; for those in between freedom is more often than not a racket.

Dr. Nathan Cohen says, "We are regressing to the notion that the individual is responsible for his actions, not his institutions," and he is heard with reverence rather than amazement. The individual—the man who *is* and *must be*, responsible for his actions—is thus debased from a creature made in the image of God to a man-made mechanism.

The gifted Englishman K. R. Minogue has written in *The Liberal Mind* that "Liberalism is goodwill turned doctrinaire." It cannot accept with grace that the primacy of "desires and needs" is not meaningful for every man, much less the free man. The liberal places desires and needs, or what he thinks are needs, at the basis of his philosophy. Satisfy needs, fulfill desires, make proper obeisance to the third goddess in this secular trinity—progress—and man's wildest dreams will be fulfilled. In the face of this obstinate disregard for reality Mr. Minogue asks, "How many visionaries have unwittingly prepared a hell on earth because their gaze was stubbornly fixed on heaven?" Perhaps the progressive educationist would be most startled to learn that "Stupidity is a moral as well as an intellectual crime." Of course Dr. Cohen would probably reply, "No men are stupid; only the teacher, or the institution, is ineffectual." Fundamentally, the urban Negro problem boils down to whether one believes that slums make the man or that man makes the slums. Certainly Dr. Cohen will never sing with the poet:

> *In your own hands*
> *The sin and the saving lies.*

It is of course possible that Dr. Cohen would not say what he did before a congregation of *his* own God-fearing people. But he was talking about people in the "ghetto" of Watts, for whom he thinks a new and special kind of ideology is required. But the Negro, more than most Americans, is a religious man, and he will someday throw Dr. Cohen's subservience back into his face. But that so horrid a collectivist and crudely deterministic philosophy can be preached to members of the American Congress, in an institution founded on an exactly opposite philosophy, is something to ponder. We must begin to grasp the magnitude of the moral breakdown in our society, in which some of the disaffected young court self-destruction through drugs and many young Negroes take to violence to attain unattainable ends.

The ends sought by some Negro extremists are indeed unattainable. Certainly the nation will never tolerate the kind of *apartheid* they preach. Nor will it allow the republic to be destroyed by a couple of "alien-minded," as distinct from alienated, agitators.

Actually, most Negroes ask for nothing more than equal opportunity to go as far as their talents and ambitions will permit. Meanwhile, social workers and politicians courting the Negro vote, professors and newspaper columnists short on copy, and television cameramen anxious for good action shots present a picture of the American Negro that is almost totally at variance with the truth of the Negro majority.

The Negro in the United States is both better off than any other white, black, yellow man in the world and worse off than the majority of Americans. Booker T. Washington once said, "When we rid ourselves of prejudice, or racial feeling and look facts in the face, we must acknowledge that, notwithstanding the cruelty and moral wrong of slavery, the ten million Negroes inhabiting this country, who themselves or whose ancestors went through the school of American slavery, are in a stronger and more hopeful condition, materially, intellectually, morally and religiously, than is true of an equal number of black people in any other portion of the globe." That statement is as true today for 20 million American Negroes as it was when it was first uttered some seventy years ago.

Yet some Negroes *do* live in slums; Negroes *do* have more broken-down apartments; Negro tenements *do* seem to be more infested with rats and pests than are other tenements; Negroes *do* have higher rates of delinquency, crime, and illegitimacy than do other people; Negro children *do* show marked inability to keep up with their white (or yellow) counterparts; Negroes *do* constitute an enormous proportion of all people on public welfare. What to do about it? Whose fault is it? Before we answer these two questions let us state the positive side of the picture.

We have already quoted from Booker T. Washington to the effect that the American Negro was in his time infinitely better off than any other black man in any part of the globe. He still is. What is more, Mr. Washington urged in his famous speech of 1895 at the Atlanta Exposition, that the Negro prepare himself for trades rather than seeking degrees. His reasoning was that if there had been a rush to obtain learning after Emancipation, the rush was understandable but, he warned, not realistic. His words fell on receptive ears.

Five years after his speech, he organized the National Negro Business League; there were then approximately 30,000 Negro businesses in the South alone. Negroes operated 900,000 farms, 1,000 millinery stores, 7,000 grocery stores, 50 banks, and 120 insurance companies. Approximately 34,000 Negroes taught in thousands of schools and colleges for Negro students. Unfortunately the young, Harvard-educated Mr. Du Bois was impatient with the slow progress of Negro advancement; in his early years he preached a moderate approach to higher goals and insisted on the importance of education as a means to *total* emancipation of Negroes. By 1905 southern black illiteracy rates compared very favorably with those of Spain (68 percent), Latin America (80 percent), and other largely agrarian countries like Russia. The American Negro rate was 44.5 percent. It was only slightly higher than that of white agrarian communities in the nation. This country was then—it is important to remember—largely an agrarian nation. By 1930 there were 15,000 Negro college graduates, forty with Ph.D. degrees, and sixty-five with Phi Beta Kappa keys.

In November 1948 *Ebony* ran an editorial entitled "Time to Count Our Blessings." The message was that it was also time to "stop singing the blues." Negroes had lots to be thankful for then and they also have lots to be thankful for now. Most of them are thankful. I speak of the hundreds of thousands of Negroes in fraternal organizations, like the Elks and the Masons, that number more living and dues-paying members than do all the so-called "civil rights" movements combined. Especially is this true of the more than 5 million blacks in the National Negro Baptist Convention.

In 1967, the year when mobs ran riot through the streets of

the United States, *Ebony* placed a full-page ad in the August 24 edition of *The New York Times* announcing to potential advertisers that 89 percent of all Negro households read its largely middle-class-oriented pages in one single year; 43 percent, or 2.5 million households, were immediate recipients of the publication, which had a circulation of 1 million; and that the median income of this relatively vast circulation was $6,648!

This figure is phenomenal, comparing well with statistics of a similar nature put out by the nation's largest mass-circulation periodicals. The approximately 90 percent reader interest shown in this moderate periodical, which invites the brand-name advertisers of the nation to share in this legitimate "black market" is a wholesome sign that, despite the sloganeering by professional radicals and misguided bleeding hearts, the Negro community—by and large—is a wholesome participant in the hopes and aspirations of the nation.

Most Negroes live very much as the rest of us do and are content to pursue happiness as it comes with whatever extras luck and natural endowment allow. Those who preach absolute equality do the Negroes and the nation a disservice, for there is absolutely no such thing. Scholars who teach or preach it do not take into account that man living freely in a democracy can be equal only before the law, the same for Stokely Carmichael as for the President of the United States. No amount of clamoring for "uhuru," for "freedom now," for "absolute social equality" will obliterate the cruel fact that not every Negro can be a Willie Mays, a Leontyne Price, or even an H. Rap Brown; no more than every white man can be an Einstein, a Kennedy, or even a Jimmy Hoffa.

It takes skill and luck to be what one wills. Many of us lack the skill, and still more of us lack the luck. We make do with what we have, always striving for more and better. But, once the notion of the better life is distorted to mean the worse life—without freedom—then the only thing to do is face the truth, refusing to evade or to equivocate or to tell polite little lies. It is the polite little lie that the Negro resents most. The scholars, against whom he is almost helpless—and who is not?—are always trying to justify their basically *useless* existence by coming up with ideological gimmicks that will attract attention, if nothing else.

America mass-produces scholars almost as it does automobiles. The 1960 U. S. Census classified more than 7 million people as professionals and intellectuals. No wonder the nation is in trouble. Fortunately there are only some 288,000 Negroes in that category, which seems just about enough—if not a few too many. The Negro who goes to college refuses for one reason or another to prepare himself for medicine, engineering, business administration, or the electronic and computing industries, which have more jobs open than applicants. To the argument that he would find employment difficult to obtain even if prepared, the answer is that no truly qualified Negro engineer, physician, or mathematician is jobless today because of his race. If he has no appropriate profession it is because he does not want one.

It is doubtful that today—with all the civil-rights laws that are on the books—huge concerns dependent on government good will and contracts would dare defy the national consensus in favor of the qualified Negro and even the Negro who is not quite as qualified as the next man. It is doubtful, I repeat, that the argument holds any water today. In any case, the N.A.A.C.P., which employs capable lawyers to look after such things, and the Urban League would soon see to it that the properly prepared Negro was hired. What is significant is that Negroes show a distaste for trade, which they perhaps inherited from young slave masters who were not as well prepared for work as were the freed slaves, with their skills. As in the family we find a deterioration, or regression, to a more permissive structure so in work the pattern of self-employment and business interest has also declined among Negroes.

The Negro rush to enterprise and the mechanical arts, to skilled trades like catering and cooking, and to insurance, banking, and even management of retail establishments has been halted; in proportion to the growth of the Negro population, there has actually been a marked decline. In 1960 fewer than 3 percent of all self-employed businessmen were drawn from the 10 percent of the population that is Negro. A similar proportion was engaged in retail businesses; Negroes operated almost 6 percent of the eating or drinking establishments but fewer than 1 percent of all other retail enterprises!

As with the Negro family, this pattern may simply reflect cultural preference. The Jewish businessman in Newark who said that he does not earn as much as a Negro does on a job or as some Negroes do on relief meant that, for the seventy-two hours a week he works in his store, he takes home little more than $150 a week for his family. He may have a point. It is generally recognized—by many Negroes too—that few people would put in so many hours for so little.

Surely the Jewish businessman or merchant does not mind filling in where the Negro refuses or has no inclination to enter. For a Jewish merchant hours are of small consideration as far as *parnosseh*—making a living for one's family—is concerned. If fewer than 1 percent of Negroes—and that is all the statistics tell us—are willing to open shops, grocery stores, delicatessens, hardware stores, television shops, or liquor stores, then the ethnic group most able to do it will. It is hardly fair to attack members of the latter as "Goldberg," "slumlords," "greedy Jews," "stingy Jews," because they as small businessmen, must charge more for the merchandise they sell; even they are unable to compete with huge supermarkets nearby.

The Negro must be clearly told that if he does not want to open a shop then someone else will; usually the someone else will be a Jew who has had much experience in the business of business. (It may be noted in passing that the Puerto Ricans are fast filling the retail vacuum in their neighborhoods from which the Jews have departed.)

Here is how Eric Lincoln, the Negro biographer of the Black Muslim movement, has put it:

> The Jew opens a business and hires his whole family. Meanwhile, the so-called Negroes [in Black Muslim parlance all Negroes are referred to as "so-called"] are footing the bill, but there isn't a black face behind a single counter in the store. . . . He will open up another business. Still later he will open a liquor store. . . . Soon he follows his Negro customer home and buys the flat he lives in. By that time the Jew is providing the Negro with his food, his clothes, his services, his home and the whiskey he has to have to keep from hating himself. But the Jew doesn't live above the business anymore. He's moved out to the suburbs and is living in the best house black money can buy.

Mr. Lincoln's indictment is nasty; all the nastier is its quotation in *Ebony* by Rabbi Richard C. Hertz, whom we have mentioned before.

Mr. Lincoln—who is neither as ill natured nor as mindless as the quotation suggests—should be reminded that nothing is stopping the Negro from owning his own clothing store to clothe himself; his own grocery to feed himself; his own liquor store to drink himself to death if he wishes; or his own apartment except his own disinclination to do so. I personally know a Negro woman who does housework for a schoolteacher friend of mine; she owns two houses in Harlem, and *she* is not "followed home" by any Jewish landlord. She owns a better home than does her Jewish employer and sends her children to college; when I read Mr. Lincoln's passage to her she laughed out loud and remarked good-humoredly: "No Jewish landlord is following me home. I'm following him to his home—to buy it, man, to buy it." James Meredith found it rather profitable (and easy) to pick up some property in the lower Bronx—and to exploit it to a degree that made some "slumlords" envious. And, in Chicago, one of the largest slum parcels in the windy city was that owned by Lorraine Hansberry's parents, the fat profits of which provided Lorraine the best education that money could buy.

Racial attacks like Mr. Lincoln's are not particularly demeaning to their Jewish targets, who can and do shrug them off; they have heard such stuff all their lives. But such attacks are profoundly demeaning to Negroes whose characters and human potentials are being subtly mocked. Who is the Negro that is fed, clothed, housed, and debauched by the Jew? Is he some simpleton, a potential recruit for one of the "action" groups whose members are *afraid* to compete in the society about them and who lash out at the Jew, the man who refuses to fight back? Does not the Jew have the right to move out? Who says that the Negro cannot do the same if he wants to? The Negro who lives in Harlem, however, has no intention of moving out, and I do not blame him. Not so long ago it was believed that the solution to his problems would be to move him from 135th Street to a high-rise apartment building somewhere else. It was the dream of Whitney Young and of all the social workers who forget that their "clients" are human beings and see them only as numbers or types in a social equation.

I think that as soon as the Negro is encouraged to go into business for himself, to be a businessman, to enjoy the fruits flowing from the cornucopia of American free enterprise without favor or prejudice; as soon as he realizes that it is not unmanly to own a grocery and stand behind a counter for many hours a day—then most of his social disabilities will disappear. One can also, however, be certain that he will receive all the help toward this goal that can reasonably be offered. What kind of help will be spelled out later. At this point I emphasize only that some Negroes *have* made it in business and that some have made it *big*.

Although still a negligible element in the total economy, Negroes have prospered as merchants, farmers, bankers, and insurance executives. The tale of how Asa T. Spaulding worked his way up to head the all-Negro North Carolina Mutual Life Insurance Company is a Horatio Alger story that many Negroes might emulate. The same can be said of John S. Stewart, President of the Mutual Savings and Loan Association, and of Norman G. Houston, head of the Golden State Mutual Life Insurance Company. These men have an answer to black power: green power.

We all know of the many artists and athletes that the Negro community has produced; they attract mixed audiences that are the best answer to the angry—and frightened—separatists who shout "black is beautiful" as if they did not really believe it. Perhaps they do not. Substitute mechanisms are a phenomenon better understood by such distinguished clinicians as Kenneth Clark and Alvin Poussaint.

I know that "integration" is a dirty word in black parlance today. But organic integration—of the kind that the Irish, Italians, Jews, and Chinese have achieved is possible for the Negro regardless what the redneck in the South or the Red in the North says. Maoism, Castroism, "blackism," or any kind of racism will not achieve it, and a black *jihad* is no more attractive a prospect than one conducted by the Egyptians against Israel. It is an ugly prospect, not fitting for a man who is *mutatis mutandis* part of Western civilization to support. The Negro should be an integral part of American civilization. He is not an African, a Chinese, or a Cuban. He is an American. If he hopes to make much mileage out of statements like "Keep your Bach, but leave our boogy-woogy alone" or if he thinks that his African heritage is stronger than his Christian American heritage, he is mistaken. Most of Negro America rejects the unwholesome notion of a separate nation.

As Ralph Ellison says, "I also inherited a group style originated by a 'black' people, but it is Negro American, not African." St. Clair Drake writes, "There is general agreement that African cultural survivals are relatively unimportant as compared with non-African elements in contemporary Negro subcultures in the United States." Except for the word "subcultures" he is generally right. I say "except" because the Negro in the United States does not belong to a "subculture," any more than does any other American, white, yellow, or brown, Jewish, Protestant, or Catholic. About the only groups that may properly be classified as having "subcultures," a term that anthropology has made fashionable, are the Amish in Pennsylvania and the Midwest and the Hasidim in Williamsburg, Brooklyn, New York. Both these groups exhibit cultural patterns so far removed from general Christian or Jewish behavior and practices as to justify the use of the term "subcultures." As for the Negro, he is no more a part of an American "subculture" than are the League of Women Voters and fans of the New York Mets.

That the Negro does not go into business with sufficient determination, does not enter certain professions, and does not seek managerial positions is only too clear. Less than ten years ago, when statistics were available, only about 1,300 Negroes owned manufacturing concerns, and they were largely engaged in producing special Negro cosmetic products for beauty and barber shops—hair lotions and straighteners, special soaps, bleaching creams, and equipment for retail shops catering to beauty services for Negroes. But disaster confronts this fragile industry, thanks to the popularity of the "natural" or African look and the proposed boycott of most such products voted at the 1967 Black Power Conference in Newark. (I understand that the entrepreneurs of this bustling black industry are not too disappointed. The fuzzy and frizzed and bushy haloes young blacks of both sexes are now sporting are very expensive to maintain. This style, by the way, owes more to the Fiji Islands than it does to Africa. But once one is intent on adopting a new image—in a hurry—one is bound to get hung up on anything that looks different enough to shock. Only nobody is shocked. Which means that the style will probably go out of fashion—soon.)

Even worse, the trend toward Negro entrepreneurship is steadily declining. "Between 1950 and 1960 the total number of Negro businesses shrank by more than twenty percent," we learn from the *American Negro Reference Book*. The decline has continued through the 1960s, and there seems to be no prospect at the moment of reversing it unless *Negro self-awareness takes hold* and a determination is made to halt it. Such a step will, of course, not be easy, any more than it will be easy to solve the problem of the Negro family. But business is a much less sensitive area than family relations, and suggestions can be more easily offered without incurring the hostile reactions that occur when other aspects of the Negro-Jewish confrontation are discussed.

To talk of "embourgeoisement" and to repeat Marxist-Maoist slogans is to serve the Negro people ill. The longer the so-called "responsible leaders" temporize and embrace "militancy" while abjuring "moderation" in order to hold the "masses," the more removed from majority Negro beliefs and aspirations they will stray. These majority beliefs and aspirations can be better understood from reading *Ebony* than from *Freedomways*, *Liberator*, or the *Black Panther*.

A recent survey by the U.S. Department of Health, Education and Welfare showed that "only" 5 percent of Negro college men become proprietors, managers, or business officials, whereas 22 percent of white college men do. Had the Department researchers crossed the street to the Department of Commerce they might have been enlightened by the discovery that Negro colleges are woefully deficient in courses in business and commerce. Yet some self-help projects have been established in the "central city." In Harlem at the beginning of 1967 about 2,000 delegates met and organized the Ad Hoc Planning Commission to organize Harlem's first community corporation. Junior Achievements, a national organization dedicated to instilling independence and enterprise in young Americans set up its first two chapters in Harlem under the auspices of the New York Telephone Company.

In Philadelphia, Reverend Leon Sullivan has had considerable success with his Q. N. program. "Qualified Negroes" is what "Q. N." stands for. He soon found, however, that there were not enough Q. N.s to fill all the genuine requests for employment he received, so he started his own program for training Negroes. It is called Opportunities Industrialization Center and is housed in an abandoned police station. He decided to instill self-respect in his clients first. He has said: "You can't train someone by just putting him behind a machine. You've got to see that he is properly motivated and has a measure of self-respect." The program is an all-around one to prepare each man not only for a job but also for his community. Training includes handling money, personal hygiene, and grooming.

"We are not interested in giving people diplomas," Reverend Sullivan says. "We are interested in giving them jobs." His efforts (according to *Time*, March 3, 1967) have been rewarding both for his "students" and for the community. The project has placed 90 percent of its graduates and has added some $9 million of purchasing power to the Philadelphia community. It is estimated that he has saved the city's welfare department $2 million in relief payments. When Federal experts came to Philadelphia they liked what they saw and by early 1967 had been instrumental in helping to set up models of the O.I.C. in eight other cities.

Of course, self-help projects are not new to the Negro community. But in the welter of publicity that surrounds so much "civil rights" activism not enough is known of real and significant attempts by responsible Negroes to help themselves and their communities. It has rightly been observed that the television camera runs to a riot but crawls to a convention.

In late August 1967, when about 3,000 Negro marines and their families met at New York's Hilton Hotel to plan guidance for returning Negro veterans from Vietnam, not a camera was there to show this dedicated group working for the Negro's and the nation's best interest. Television executives probably believe a riot makes a "better" picture than a meeting does—but it is a distorted picture all the same. Former CORE chief Floyd McKissick remarks:

> Today there are only two kinds of statements a black man can make and expect that the white press will report. First is an attack on another black man calling him an Uncle Tom, or a fanatic or a black nationalist. The second is a statement that sounds radical, violent, extreme—the verbal equivalent of a riot—Watts put into words. . . . Think back over the past months. You will begin to realize that the Negro is being rewarded by the public media only if he turns on another Negro and uses his tongue as a switch blade, or only if he starts a riot. . . . How many of you report even what middleclass Negroes do? . . . We'd like to feel what we did on the local scene was important. You know, we like news clippings, too.

How terrible an indictment, yet how true!

The Harlem Ad Hoc Committee went about its efforts to help build a middle-income apartment house in the community that had been thwarted for a half-dozen years; the U. S. Housing and Redevelopment Board, which has other notions of integration has, however, turned a deaf ear to its demands. As committee members are not a sitting-in, lying-in, or window-smashing segment of the community, they receive little or no publicity from the press, radio, and television. But in 1969, when a small but determined group of black rebels occupied a Harlem building site selected by the Governor of the state for a 20-story office building, all construction was stopped. Later a rump meeting of unrepresentative militants (including a couple of local politicians) "voted" their power to determine the life or death of the project. At this meeting Vincent S. Baker, a vice-president of the NAACP, was brutally manhandled when he rose to speak in support of the Governor's plan. Said Baker: "It was the closest thing to the rise of Nazism in this country since the nineteen twenties."

On February 11, 1967, *The Amsterdam News* featured the twelve-year-old Morningside Citizens CARE Committee in a three-column story with a three-column picture of some committee leaders. "CARE" is an acronym—without acronyms many ethnic organizations would really collapse, it seems—for "Clean Areas Reflect Efficiency." I doubt that any of the six people in the picture had ever been televised or interviewed before. They had all been trying—even before massive Federal funds were lavished on the community—to do a job on their own, and they had largely been ignored by the television cameras. A singularly effective aspect of their work is the aid they give to property owners—the dread landlords—"whom they help in improving their houses and carrying on neighborhood conservation."

The Morningside Village Center (a subsidiary of the committee) conducts an after-school program for children six to thirteen years old, a play center for preschool children, and a senior-citizens center. It does so without subsidy. I do not think that the Federal government, the state, or the city has ever given it a dime. The same is true of the Negro boys and girls engaged in Junior Achievements. They have never been on television, yet any black fanatic can command a nationwide press conference, abusing the white press in language that makes one wonder why white reporters and network television put up with it. Is it possible that the angry black nationalist who accused James Baldwin of "copping out" because he refused to publish in *Liberator* after it had printed some anti-Semitic articles has a point when he says that "whitey . . . wants to be whipped"?

Perhaps the photogenic boys who broadcast the television news do really want to be "whipped." I know that I do not, nor does the majority of the American people. But television ignores the decent public as it caters more and more to tastes for violence. The boys and girls and the men and women of the law-abiding and patriotic Negro community have as much chance of being seen on television as programs of taste, delicacy, and refinement.

It is disquieting to learn from a report prepared by the Institute of Defense Analysis for the National Crime Commission that nine of ten urban Negro boys face arrest at some time in their lives (traffic violations excluded). If that is so, and we have no reason to doubt it, what have all the gains of the civil-rights movement amounted to? Is it possible that young Negro boys—fatherless and socially aggressive—court arrest through vandalism and petty theft?

A young hoodlum was questioned by a television reporter and asked why he had participated in a riot in Jersey City; he replied: "Why? Because I'm culturally deprived and I'm from a broken family that is matriarchally dominated." Ralph Ellison, who detests the cant of the sociologist, reported this answer, adding, "The reporter was just getting the feedback from the sociological jargon that is going around." "Sociological jargon" is jargon loaded with dynamite. It encourages a breakdown in the moral climate, as sociologists find new labels for old delinquencies. The wisdom of that boy is, however, not recognized; only the sermonizing and preaching of the television editorializers are. Meanwhile, the crime rate does climb in the black slums, as acts of violence and just plain murder come to be regarded as part of the Negro's struggle for civil rights.

When Whitney Young, Jr., says that 99.9 percent of the Negro people cannot tell a Molotov cocktail from a martini, he is probably right. And, when he says that 97 percent of all Negroes have never participated in a riot he is again telling what seems the unvarnished truth. But, when he implies that poverty causes riots, that bad housing causes fire-bombing, that lack of jobs is cause for looting, that the "ghetto" causes Negro unrest and rioting, he is ignoring the facts. Certainly more than .1 percent of all Negroes are poor, ill housed, or jobless, and certainly more than 3 percent live in "ghettos." If the causes were that simple, the cures would be easy enough to find. Surely Mr. Young knows that hundreds of billions of dollars are not necessary to solve the problems of fewer than 3 percent nor hundreds of millions for fewer than the one-tenth of one percent.

Actually, Mr. Young is saying two things at once. On one hand he argues that only a very small proportion of the Negro urban population that lives in "poverty" reacts to this condition with violence and socially destructive behavior; on the other, he claims that the vast majority, living under the same conditions, does not react in the same way. His argument is comparable to a hypothetical National Health Service report that excessive cigarette smoking causes lung cancer, except that among all heavy smokers 97 percent show no carcinogenic symptoms at all. Such a report would be laughed out of the market by the tobacco companies, and with cause. If something is a cause it should affect more than 3 percent of the people to whom it is applied. Not that a one-to-one correlation is necessary, but 30 percent or even 13 percent would be a more convincing figure.

It must be emphasized that much black delinquency in urban centers is rooted in a unique family situation that no amount of Federal money for housing, jobs, education, and so on, will help. There is no question about this problem, and no amount of arrogance or polite sociological jabber will do much to alleviate it. Daniel Moynihan urges a guaranteed annual income to improve Negro family life in the slums. His proposal is tempting and certainly more reasonable than most such proposals by politicians and reformers, but he is woefully mistaken if he thinks that it will find acceptance among even the most moderate Negro leaders. The Negro is not automatically against any proposal from a white man, but this one is self-defeating, and the Negro knows it better than Mr. Moynihan does.

When Mr. Moynihan appeared before a Senate committee to argue for his plan, he was opposed by Kenneth Clark, who called it an attack on the "dignity" of the Negro. Mr. Moynihan argued that guaranteeing every Negro man a job and every family an income would tend to keep the man at home and the family together. Dr. Clark, who knows the Negro family at close hand and who does not perhaps share Mr. Moynihan's "optimism" would have none of it. Dr. Clark reminded the committee that "Pouring money into slum schools makes very little difference."

He added, "Education is not something that principally takes place in schools." Mr. Moynihan might have replied that the family is largely responsible for the pupil's attitude toward school, but he did not. Clark went on to dispel some dearly held illusions of those (Mr. Moynihan among them) who believe that enough money can cure almost all the ills society is heir to: "[we] can't use the pathology of the slums as an excuse for the underachievement of Negro children in the schools." Like most Negro leaders, he made obeisance to the goddess poverty: "The possibility for dignity simply does not exist in an environment of stark material deprivation." Now this supposed "material" deprivation runs counter to researches, statistics, and reports of actual "material" possessions of Negroes, especially in the urban slums. Certainly it does not fit with the statement that for 1960–1961 "the average urban Negro family spent $3,707 for annual living expenses." This amount, according to Joseph H. Douglas in *The Urban Negro Family*, "was just over two-thirds that spent by the average family in the nation as a whole—a ratio which was virtually unchanged from 1950." This figure hardly adds up to "stark material deprivation"! Nor does it reflect a loss of "dignity" that cannot be regained unless the Negro family spends almost $6,000 a year. There may be no second television set, no color television, no new-model car, no two-bathroom apartment, but the income does not spell "stark" deprivation. Almost *all* Negro families in Harlem, in Watts, in Cleveland, in Newark, in Detroit, in New Haven—in the cities where the biggest riots have occurred in recent years—possess television sets, cars, bathrooms, hot and cold running water, steam heat, and electric refrigerators, "material" possessions that the average European worker does not yet enjoy. If lack of material possessions were cause for riots European workers would riot. Russians who have fewer possessions than the poorest family in Harlem would be in for such a round of riots and disturbances as no amount of police repression could quell.

If the "average" Negro family in the city spent about $80 a week a decade ago—today, it spends more—it could fulfill, even at prevailing inflated prices, many, if not all of its needs for food, clothing, and shelter with money to spare. According to a very recent (mid-1967) survey, the average American family (Negroes included) spends about 20 percent of the family budget on food. The European family spends 30 to 35 percent and the Russian family about 50 percent. The American family—the Negro family included—thus has a greater margin of his income to spend for other things.

The Negro is not much different from other members of the great American family; certainly his possessions, his "material" gadgets, are probably *more—not fewer*—than those of other, white minority groups in the nation. To speak of "stark" deprivation, of the lack of "material possessions," of the "environment" in such harsh terms is to imply for material things a power over Negroes that they lack among almost all other groups. Actually, Dr. Clark seems to be saying that *Negroes*—of all the people in the land—cannot have "dignity" because they are poor. He not only slights the poor; he also gratuitously insults the Negro community.

Should Negroes in our urban centers therefore be contented to demand no more than what they already have? Of course not. Negroes for the most part dream the same dreams that white Americans dream. Realizing the dreams is harder for Negroes for a number of reasons. Education, family structure, social attitudes derived from them, and a host of other factors make the Negro problem a special problem. But it is not so special as to require that the whole nation go on a binge of "reform," some of it motivated by desires to capture the Negro vote. If the "urbanologists" of today think that tomorrow's projects to correct urban blight will contribute to dignity, education, and stable family structure, he is woefully mistaken. The nation will pay a higher price tomorrow than it has paid to date unless it learns from the past.

When the social reformers thought up the high-rise monstrosities that disfigure our cities only about a generation ago, they sat back pleased that at last a start had been made in ridding urban centers of the "worst features of the slums, overcrowding," as one of them put it. Today these oversized publicly endowed institutions for hopelessness and discontent, these dens of delinquency, are recognized as the blight they are, yet no one, least of all those who thought them, up, apologizes for the *grande gaffe*. Instead new projects are planned; they are to be publicly endowed with "private" interests sharing in the process in hopes of avoiding the stigma of previously all-public failures. But delinquency, illegitimacy, and learning failures among young and poor Negroes have not been alleviated one single bit.

In fact, misery has only been intensified. Should slums not be cleaned up? Should the nation ignore urban blight? Should we simply not give a damn? Of course not. But rehousing all Negroes will not modify seriously their lack of orientation to education; will not prevent future riots; will not convert the Negro family into the stable institution that Mr. Moynihan and others would like; will not instill dignity, love of truth, or freedom.

Worse, it will only help exacerbate already tense relations because dignity, love of truth, and freedom cannot be *forced* on people. There are some things that money cannot buy. Slum clearance should be a major concern of the community. The Italian on Mulberry Street in New York, the Polish resident of the Ironbound section of Newark, or the Irishman in Milwaukee may want his neighborhood to remain as it has been for the better part of a century. He is not rioting because of it, and he will thank national planners and politicians to stay the hell away from his "slums" if they want his vote. Knowing that such an appeal will fall on deaf ears, politicians never offer to clean up *these* slums, and they are not much better externally than any in the Negro areas of the nation—*and a good deal worse than most.* But in the belief that some politician in the urban Negro community speaks for all its residents, energetic politicians think to make themselves popular by calling for instant housing and urban renewal—only to be caught up short by young black power militants who know that, when Negroes are dispersed all over the city, they will not even be able to elect an Adam Clayton Powell. The militant cries, "Renew somewhere else, and leave *my* slum alone." But, knowing that he might have to find an honest job, the Negro militant rants about rats, ghettos, deprivation, and whitey; he burns down a few stores or threatens to level the city, all the time knowing that white politicians are interested only in his vote. If the politician is also a liberal and well-meaning white man, the black militant will find it easier to con him into a $50-a-day job—helping keep the peace—than to help eradicate "poverty."

At the root of the poverty problem in our Negro slums, at the root of the unrest in the better-placed Negro communities, and at the root of much of the unrest that emanates from angry young Negroes is shame at the decision of the U.S. Supreme Court, which singled out Negroes as the only minority in the land incapable of acquiring education when left to themselves. We shall develop this theme in the next chapter, but here we shall argue only that much of the so-called "Negro protest" is generated by fear that perhaps he cannot make it in a white man's world. Another contributing factor is a kind of radicalism of the *lumpen-proletariat,* combined with a kind of nihilism preached by representatives of the *lumpen* middle class. The latter, betrayed by their Marxist faith, turn to variations preached by Mao Tse-tung, Frantz Fanon, Che Guevara and Herbert Marcuse. They preach violence for its own sake, to which they add a simple black racism, which has more appeal for the black masses than social thinkers had previously thought possible.

In any case, the new black insurgents are saying: "Don't rehabilitate *our* areas; don't plan for urban renewal in *our* neighborhoods; don't improve *our* blight." We may as well listen to them. Massive programs to resettle, relocate, and rehouse Negroes living in the so-called "ghettos" are not *now* Negro demands. They are the solutions sought by white liberals who want to prevent the fire next time. James Farmer is outspoken in opposition to urban disposal; Claude Brown, who speaks with perhaps more anger and more understanding of the demands of younger Negroes in slums and colleges, says that what the Negro wants now is *not* what the liberal offers him. What he thinks the Negro wants now is "instant freedom and instant social equality."

Whatever Brown means by the latter term, and he did not make it clear in a television appearance in August 1967 on Channel 13 in New York, we can infer that he wants to be recognized as equal with the white man, not as someone *special* to be despised or catered to. That is how I understand his remarks. I find nothing either frightening or frivolous in them. That it is possible to achieve his goals instantly in the United State I doubt. But they are certainly as valid goals as are those of massive aid and one-sided programs to *help* the Negro. Because the latter programs have built into them a degrading element of patronage for an inferior group, they are being rejected by young Negroes. If we can dissuade these Negroes from a blind Marxism and demonstrate that Soviet Russia is no friendlier to the Negro than to other minorities it cannot absolutely subjugate, that Mao Tse-tung has no interest in the Negro plight beyond exploiting violence to bolster his foreign policy, that Castro has few truly black men at the helm of his regime, that Communism is the most antipathetical ideology to American Negro traditions—then perhaps it may be possible to renew the meaningful dialogue that was rudely interrupted by *Brown v. Board of Education* in 1954.

First however, we must understand that a few years ago it was believed that Negro riots would be cathartic. We have since been told by Charles Silberman, author of *Crisis in Black and White*, that they are "cumulative."

Nor are riots confined to those areas where, as Kenneth Clark says, "stark material deprivation" prevails. On the contrary. What the riots have proved incontrovertibly is that people deprived in the way Dr. Clark describes are nonexistent or that those better situated are as much fuel for the fire as those who are not. The Watts experience, where most of the rioters had high-school diplomas; the riots in Cincinnati's Avondale and Hawthorne suburbs, which represent almost ideal living communities; those in Nyack, New York, and Waterloo, Iowa, where Negro unemployment is negligible; those in "affluent" Plainfield, New Jersey; those in relatively high-employment areas like Detroit or in Federal-aid cities like New Haven all suggest factors not usually stressed in discussing the "cumulative" effects of the riots. In fact, the Kerner Commission Report stresses the relative "opulence" and "better education" of the average rioter. Among those charged with the most extreme offenses to the public order in New York are black men high up the ladder of professional success; in California the head of the most militant organization of black *schrecklichkeit* is Ron Karenga, a graduate of the University of California who holds a master's degree in political science and who is so far out in his demand for "black culture" that he insists on being called "maulana," meaning master teacher. We have supplied sufficient illustrations of riots in which the leaders are not "starkly deprived" or lacking in "dignity." It is the sons of the black middle class who are most disaffected. This phenomenon is understandable, for it follows a recognizable pattern among leaders of violent revolution from Robespierre to Lenin, from Lenin to Castro, from Castro to Carmichael and Karenga. (The white students who riot on the Berkeley and Columbia campuses are all of middle and upper middle class families.)

It does not follow the pattern of stark poverty that the white liberal claims, and it cannot be eliminated by providing more food stamps, more housing, and a few more summer jobs or by buying off recalcitrant hoodlums in the nation's Harlems. Peace is not purchased that cheaply. It has to come through a process of understanding on both sides.

part 3

The Contemporary Scene

Chapter Thirteen

EDUCATION: INTEGRATION AND SEGREGATION

It has been said that every myth has its truth. One of the abiding myths in the United States is that given the opportunity everyone will grasp it *equally*. The primary myth, however, is that education is so integral a part of the American way of life that everybody should receive as much of it as possible, regardless of its content. Education *per se*—a diploma, a degree—is regarded as the right of every free-born American.

Some people have even argued that it is a *democratic* right. If every boy or girl cannot acquire a Ph.D. degree in physics, classical archaeology, Latin, Greek, or Hebrew, at least he should be able to acquire one in packaging, retailing, hotel greeting, or football-helmet crushability (the latter is a real, not fictitious, degree). This "democratic" approach to education fostered by men specializing in such "disciplines" as educationism—the teaching of teaching—has produced swarms of graduates from thousands of colleges and universities. It has yet to be proved, however, that we have a higher proportion of genuinely *educated* people than do France, England, Sweden, and Denmark, where higher education is limited to about 5 percent of those attending secondary school.

Men have long sought the reasons that some have done better than others in a land where everyone is "born equal," which all too often is misinterpreted to mean that we all have equal abilities if only given equal opportunities.

When it was found over the years that Negroes did not do as well as whites in army intelligence tests, in school work, or in college attendance, those interested in exploring the problem turned to the *facilities* provided both races in the elementary schools of the nation. Enough statistics were available to suggest that mere updating of physical structures or changes in curriculum would not quite meet the problem. It was decided to seek the one absolute factor that might account for the gross disparity between Negro and white achievement.

In the South it was found in the almost total separation of the races at all levels of education. The South thus seemed a logical place to attack the problem, for many northern Negroes came from the South, where they had already been shaped by their early education. It was decided to test the time-honored and oft-sustained Supreme Court decision that "separate but equal" fulfilled the Constitutional requirement for "free and equal" as embodied in the Bill of Rights.

In the spring of 1954 the United States Supreme Court unanimously ruled in *Brown* v. *Board of Education* that separate (which it called "segregated") education of the races was "inherently" inferior education. It ordered schools in the South—and everywhere in the land—to begin with "deliberate speed" to desegregate their facilities so that Negro children could attend schools with white children. That decision applied only to the elementary grades. In a subsequent decision it was extended to higher education as well. *Brown* was the greatest liberal victory since Franklin D. Roosevelt's New Deal legislation in the 1930s. (Parenthetically, it may be noted that President Roosevelt sanctioned internment of Japanese-Americans in concentration camps during World War II.) There was to be "equal education" for Negro children in southern schools, where little black children would sit next to the little white descendants of Pitchfork Ben Tillman, Wade Hampton, and Tom Watson. The whites of Dixieland shuddered at the prospect.

But the decision was made; the appeals in local, state, and federal courts were exhausted. There was nothing to do but comply—but not in a hurry. The Court had said "with deliberate speed." The years dragged on, and so did the slow process of desegregation. Meanwhile the Court continued to strike down previously established Southern folkways like separate toilets in public places. Where the Court's school decision produced the most unexpected turbulence was in the North. The northern liberal, secure in the belief that *he* would never sanction segregation in the North, was confronted by Negroes who attacked the cherished concept of the neighborhood school, until then the backbone of the free public-school system in the North.

Soon the cry for "integration" of the school system, which had never been directly challenged in the North, was changed to a cry for "racial balance" in the schools. But one man's balance is another man's saturation, and soon white liberal parents—those usually loudest in their protestations for "equal rights for the Negro" in the South—decided to move away from their school districts to neighborhoods where the chances that the Negro population would overwhelm the whites seemed more remote.

Some of them sent their children to private or parochial schools; Jewish day schools, which until 1954 had had negligible attendance, suddenly showed marked growth, as professional and liberal Jews asserted piously that the upswing showed a revival of interest in the traditions of Judaism. But they did not fool the Negroes, especially those Negroes closest to the Jewish scene—the Negro intellectuals. Soon white liberal housewives, who would previously rather have been found dead than say an unkind word about a Negro, would become almost incoherent with rage and fear at proposals to alter the composition of neighborhood schools, to require busing of Negro children to "their" schools or busing of their own children to other districts. Suddenly almost everybody realized that northern schools had become *more* segregated, sometimes to an absolute extent that challenged comparison with pre-1954 schools in the Deep South.

The causes were obvious. As more and more Negro children were enrolled in schools, previously integrated organically, almost all-white, more and more white parents took their children out of school by the simple process of moving away, making room for more and more Negro children. By 1967 the overwhelming majority of children in public elementary schools in Manhattan and majorities in the other boroughs of New York City were Negroes and Puerto Ricans. Washington, D.C., schools, which were about 50 percent segregated before the Supreme Court decision, are almost 100 percent black today. Similar trends are visible in almost all major urban areas in the nation.

But the real problem still remained. The schools had been "desegregated" in the North and, to an appreciable degree, in the South. A question that liberals had refused to ask themselves was beginning to haunt Negro leaders: Why would the child of a southern Negro receive a better education simply because he sits next to a white child in school?

"Separate educational facilities are *inherently* unequal." In an effort to explain this strange bit of racist reasoning, the Court asked itself the following question: "Does segregation in public schools solely on the basis of race, even when the physical and other 'tangible' factors may be equal, deprive the children of the minority group of equal educational opportunities?" And the Court replied, "It does."

Why? The Court decided not to elaborate but simply to let the assertion rest. The U.S. Supreme Court thus ruled in 1954 that Negro children in a Negro classroom in a Negro schoolhouse could receive only an "unequal" education because the separation from white children alone—*per se*—guarantees poorer education even if "tangible" factors like plant, equipment, and staff are "equal."

This shocking bit of sociology, which had nothing to do with law at all, was borrowed from Mr. Gunnar Myrdal, whose book, *The American Dilemma*, was reported to have greatly influenced the Court. When the true implications of the decision became clear to the mass of Negro people, it dawned upon them that a school with an all-Negro faculty of the best scientists, writers, musicians, business consultants, athletes, and educators in the land would still be inferior and "inherently unequal" if white children were not in attendance "so that some hidden wisdom would rub off," as one Negro militant remarked recently.

The reaction to this decision—hailed by all liberal elements in the land—was slow in surfacing. Most Negro leaders hailed the decision as one of landmark proportions. Psychologist Kenneth Clark even called it the triumph of color-blindness. There was celebration. The walls of segregation had finally come tumbling down, thanks to the Joshuas of the Supreme Court. The nation waited for the schools to move. And move they did, especially in the North. But, as time went on and Negro children were not found to be appreciably "smarter" than they had been before the landmark decision; as almost all urban schools became *more* instead of *less* segregated; as tensions mounted over who was to blame, other opinions began to be heard even from such citadels of the liberal establishment as the *Washington Post:*

> Some Negro intellectuals now are even beginning to question the idea that desegregation is necessary to improve the Negro condition. That view had been taken for granted for years within the civil-rights movement; it has been accepted by the courts, and underlies all the great school decisions.

But the questioning was much sharper than the *Post* indicated, as became evident after President Johnson's Commission on Civil Rights issued its annual report, which seemed to show that old hopes had died in the face of reality. The Commission found in early 1967 that, in the urban centers of the nation where the majority of the population now resides, approximately 75 percent of all Negro elementary-school students were in schools that were nearly all Negro in population; 83 percent of white elementary-school students were in all-white schools.

It was found that, as soon as a school tries to speed up an organic process of integration, it turns into a completely Negro school. This change in turn produces such parental conflicts as had never before been witnessed as a factor in American community life. In every mixed neighborhood, teachers and boards of education are on the defensive because of a decision that reflected their own ideas in the recent past. And no one has yet proved that a single Negro child is any better off today than he was the day before the Supreme Court ruled in *Brown.* But the contrary can be proved only too easily.

Our intention, however, is to demonstrate only that forcing issues almost never produces desirable results, especially when it requires placing children on the front lines to prove a hard sociological thesis. Force and freedom are *not* compatible concepts, especially in molding a child's mind. No wonder that the Court decision can be interpreted as showing *inherent* contempt for Negro abilities. Floyd McKissick finally got so fed up with the notion that he came to regard the decision as having meaning only for white liberals. "Mix Negroes with Negroes and you get stupidity" is the way he came to interpret the Supreme Court decision.

Of course, when the ordinary Negro mother heard of the Court decision on her child's behalf, she believed that the issue would be simple. As Nathan Glazer quotes one such mother, "the children go to school with other Negro children, their education must be worse." That is how she *understood* the decision. When her son or daughter began to attend mixed classes and still could not read on the white schoolmates' level, her frustration must have been overwhelming. She is no sociologist, no lawyer, no politician; she is not concerned with abstract rights, black power, or any of the slogans for which young men are willing to tear the nation apart.

At the end of May 1967 Herbert Ottley, leader of Youth in Action, a Bedford-Stuyvesant community organization to combat poverty, told a gathering of leading Negro educators that "school integration was no longer an important goal because nowhere in this world is anybody going to take care of your kids better than you." He asked that emphasis be put on education, not on transfer gimmicks that would shift Negro children to white schools and white children to Negro schools. The meeting, sponsored by the Negro Teachers Association, agreed "that Negroes could study together and did not need to sit beside white students." Dr. Alvin Poussaint of Tufts Medical College added, "There is no reason why we should have to join the white community to achieve our goals in education."

Others more outspoken wanted "Black schools for black children," and by late 1967 the New York City Board of Education had partially dispensed with its merit system for promoting principals and had decided to institute a new scheme for appointing a number of new "demonstration-school principals." The only requirement that the Board has not been able to overlook is the color of the applicant's skin! He must be black—or else—the Superintendent of Schools was warned. He heard the message and announced appointment of a half-dozen black "demonstration principals."

An old conservative concept that the less centralized power the better, was severely tested when three experimental school districts were set up in New York City: one in Harlem, comprising four schools, called the Intermediate School 201 complex; one in the lower East Side of Manhattan, consisting of six schools, called the Two Bridges project; and one, made up of eight schools that has gained renown as the Ocean Hill-Brownsville project. The I.S. experiment was the first to open. There, the confrontation was immediate. Black extremists at first denounced the school system for building "a windowless trap for black children." I.S. 201, the central school in the experiment, was so characterized because, though built at a cost of five million dollars, it was "windowless." However, it had no windows because it was the first public school in the world that was completely air-conditioned. Its other equipment was equally advanced.

At I.S. 201, the local governing board appointed as one of its top aides Herman B. Ferguson of the Revolutionary Action Movement, an organization charged with conspiring to assassinate a group of civil rights leaders including Roy Wilkins of the N.A.A.C.P. Though a confrontation had been smouldering at I.S. 201 since 1966, it was the confrontation in Ocean Hill-Brownsville that was the one that would set the city on edge. A crisis in decentralized education was exacerbated by two forces—each with its own power concepts, each insistent on total victory, each convinced of the justice of its cause. On one side in the struggle was the United Federation of Teachers, headed by Albert Shanker; on the other side was the local governing board, headed by the Reverend C. Herbert Oliver and the local district administrator Rhody McCoy. Junior High School 271 became the focal point of the dispute when some ten teachers, all union members, were dismissed in May, 1968. Some three hundred teachers in the local district went on strike. Meanwhile a hearing was held before a trial examiner—a Negro jurist—who dismissed the charges leveled against the striking teachers by the local governing board, but they were not readmitted. Local extremists brought considerable pressure to bear on the board and on Mr. McCoy not to permit the return of the union teachers. At this point Shanker warned the central board of education that he would strike the entire school system when the fall term opened in September if the teachers were not reinstated. On September 9 school opened, but the striking teachers were not allowed to enter the district schools. The confrontation became increasingly ugly as the union and the governing board exchanged charges. Leaflets appeared, distributed by an *ad hoc* committee, calling Shanker, the teachers union and the returning teachers "dirty Jews," among other nasty names.

Most New York teachers are Jewish. For this—and because the majority of Negro pupils are behind in reading skills by at least two years for reasons a thousand studies and reports have not yet made clear—certain black demagogues accused the teachers of committing "genocide" on black children. The bitterness aroused by the Ocean Hill controversy is not yet quieted down, even if the State legislature has passed a bill requiring that all New York City's school districts contain no fewer than 20,000 pupils. This, it was thought, would effectively eliminate the smaller demonstration districts by absorption into larger decentralized groupings. But the issue will not down, law or no law, and the conflict still awaits solution.

And, fourteen years after the famous Court decision, Negro children at Swan Quarter, North Carolina were echoing a chant heard in 1954 from the throats of white children in front of Little Rock High School: "Two, four, six, eight, *We* Don't Wanna Integrate."

The hasty decision to right wrongs not properly understood by the Court has thus come full circle. Will Mr. McKissick urge the repeal of *Brown v. Board of Education* in the light of his present thinking on the subject? Will the younger black militants demand that the decision be abrogated because it holds the Negro up to ridicule? The questions are rhetorical, of course, but it does not hurt to ask them. Meanwhile, in late 1969, CORE was preparing to challenge the new integrationist push of a new administration in Washington and an almost new court by setting up separate (but equal) black schools in the South.

The Negro periodical *Phylon* (second quarter 1954), asked, "Where were the shouts of victory, the parades, the fireworks?" They were all taking place in the editorial rooms of the larger newspapers in the land. Negroes seemed to sense that the decision boded no good at all, for it posed a challenge that he was not yet prepared to face, that he may not yet be prepared to face. It is the challenge that pits him against the white man in an area that frightens him. On returning from a journey of discovery in Africa Mr. McKissick informed a press conference that he had learned that "You don't need extensive education to acquire skills . . . like all this jive we've heard for 400 years." *The Amsterdam News* (August 26, 1967) reported that "He cited Tanzania as an example." One probably does not need extensive education if Tanzania is his model. But why should it not be? Should Mr. McKissick take Periclean Greece as paradigm for the ideal life? or Renaissance Italy? or western European civilization? They are difficult models, and Mr. McKissick has the right to cop out of our civilization. But before he makes a final decision about Negro capability or the black potential in America, the educator—*not the educationist*—should have a whack at it.

The myths on which the Myrdal book was based were largely discussed in banalities and generalities that would today be laughed at if presented for serious consideration. But the book has already had its pernicious effects. No book has the power to corrupt an entire court. Myrdal's pleading approach, couched in platitudes and dripping *weltschmerz* had engulfed the so-called "intellectual" climate of the time, and anyone who dared to challenge its unidentified quotations or its pious absurdities was denounced as ignorant or unfeeling. How this book could take in an entire generation is something I shall long ponder, along with the success of *Anthony Adverse*.

Had the Court really been interested in *education* of Negro children—and not the sociological theories of academicians—it might have studied the researches of another scholar, a Negro scholar at that. Carter G. Woodson in his *Education of the Negro Prior to 1861* had observed that in early New York "much had already been done to enlighten the Negroes through schools of the Manumission Society." But, as the Negro population increased, he went on, it was decided to merge the Negro schools with the free public schools. "Despite the argument of some that the two systems should be kept separate, the property and schools of the Manumission Society were transferred to the New Public School Society in 1834. *Thereafter the schools did not do as well as they had done before*" (emphasis mine). Whatever the reasons—and there were many—the fact is that they did not do as well as when they were separate but equal. Dr. Woodson noted similar trends in cities like Cincinnati, Columbus, and Cleveland. The trouble with the "separate but equal" doctrine was that it was never elevated to "separate but better," especially in the South.

Given the Negro family and the absence of a tradition that regards learning as good in itself, the question may be asked: Could the separate Negro school have been equally good—or better? Why not? Not all education need be made available to those who are uninterested. For instance, education can serve a broad spectrum of interests, preparing students for viable jobs, etc. But if deflected to taking courses in "soul music," "soul food," or "black chemistry" (the making of Molotov cocktails, as it has been defined) then only an explosion can be expected.

Actually, when Mr. McKissick complains that his mind has been cluttered by 400 years of useless "jive," he really means that the American Negro has been overeducated. Maybe he is right. Yet when we say "the American Negro" whom do we have in mind? For there is no such thing as "the American Negro" any more than there is "the American Jew" or the "American." It is possible to create a composite, and statistics, sickening though they be since they can be employed to prove any demagogue's theorem, do have to be taken into consideration; the composite of the

American Negro is that of an attractive personality exploited by black and white leaders who almost never mean him well. Not *never*—but almost never.

In matters of education, it is paradoxical that those who have the most to offer Negroes are Negroes themselves. Unhappily the courts have been too ready to listen to well-meaning white educationists, men with all kinds of degrees who know little enough about subject matter and even less about the Negro. Booker T. Washington, who knew best how to approach this problem has probably been dismissed more often by white liberals than by informed Negroes. The same can be said for W. E. B. Du Bois, whose theories on Negro education were certainly sound. In July 1935 he wrote in the *Journal of Education:* ". . . theoretically, the Negro needs neither segregated schools nor mixed schools. What he needs is education. What he must remember is that there is no magic, either in mixed or segregated schools."

This statement was, of course, contradicted in letter and spirit by the Supreme Court decision of 1954, which insisted that separation of the races in the schools *per se* produces inferior education. Sociology thus led astray the nine wise men of the Court; a decision from which the nation may never recover.

Here it may be said that the appointment of Thurgood Marshall to the Court may prove to be of enormous benefit to the Negro, not because he himself is black but because he knows from his own experience that what he has achieved, other Negroes may achieve also. Perhaps the color of the hand that throws a Molotov cocktail will mean *less* to him than to more radical-minded jurists.

What Du Bois had in mind can be summed up: One can learn anywhere if the will to learn is there. But is the will there? In too many American Negroes it is not; others expect more from education than it can fulfill. They are encouraged by public and private propaganda extolling the virtues of the high-school diploma, which will supposedly catapult them into "better-paying jobs." How genuinely frustrating such appeals may be for the uninformed and untrained can be gauged by the number of so-called "vocational" high schools in the United States that persist in issuing diplomas for trades that no longer exist as viable sources of income in the national economy. Many are about as profitable as blacksmithing. A Negro boy with a diploma from a Printing High School or a Garment-Cutting School will be turned away if he cannot read or cut to the specifications that the job requires. His diploma will be of no moment to the employer. When he is turned down, he "knows" that it is because his skin is dark, rather than that he lacks skill. No one has told him that the diploma is supposed to represent skills acquired or knowledge gained. In the permissive environment of American urban schools he has been coddled, for his color rather than his ability, and crippled by the educational system.

Boards of education in New York and elsewhere try to devise new patterns of education to take the place of yesterday's fashions; they improvise at the cost of hundreds of millions of dollars as each year more and more Negro boys and girls emerge less prepared and less educated than before the Court's decision.

How much experimentation does the Negro have to put up with? Treated as material for experimentation, measured, tested, rewarded, and chastised according to the latest educational fashion, the young black man and woman are finally shouting: "Enough! Get off our backs. Let us, please, try to educate ourselves."

Certainly the nice young schoolteacher who always gave a passing mark to the "disadvantaged" Negro boy, regardless of whether or not he deserved it, did not realize that he would someday lash out at her. She could not understand James Baldwin's saying, "There is no place for the white liberal (in the civil-rights movement)." Baldwin also said, "Any Negro born in this country who accepts American education at face value turns into a madman—*has* to" (Baldwin's emphasis).

Why does he have to? Mr. Baldwin might have recalled how he had been treated by white teachers at DeWitt Clinton High School in New York City with a color-blindness that he later came to reject. He was made editor of the school's literary magazine, as was his successor Paddy Chayefsky, because, as his English teacher has since written, she never really looked into his pigmentation but into his mind; she found it fresh and original.

But that is Mr. Baldwin's problem. His general resentment can be understood only in terms of betrayal of something he does not quite understand. No one has ever suggested that James Baldwin is a great thinker. It is enough that he is an eloquent, if somewhat overwrought, writer. The depth of his anger is probably caused by a sense that he is being patronized by some Jewish intellectuals who have not accepted that he means *them* when he says "white liberals" have no place in the movement. In one discussion with several Jewish liberals he demanded that "ethics" be dropped from the discussion; Norman Podhoretz of *Commentary*, the moderator, dutifully remarked, "It's unfortunate that the word ethics came into the discussion." Professor Sidney Hook replied, "That's a very peculiar observation." Most peculiar. The discussion had suddenly become a parody of Lewis Carroll. Mr. Baldwin, seeing that these Jewish intellectuals lacked the dignity to defend their own values probably felt that they were simply being deferential to him; the one thing the liberal cannot fathom is that a Negro would rather be called a "nigger" than a "colored man," a "bum" or a "thief" than "disadvantaged" or "culturally deprived."

In the same discussion Dr. Hook remarked, "I consider a program which would lower the standards of *achievement* for Negroes as tantamount to regarding them as second-class citizens" (his emphasis). Yet that is just what our urban schools have been doing for a generation. It seems that some ignorant and unsophisticated Negroes are filled with outrage at the white teacher—in New York the Jewish teacher—for keeping their children at a disadvantage. Why else, the Negro mother asks, are her children two or more years behind white children in the same school? James Bryant Conant, former President of Harvard University and a distinguished educator, has stated, "What a school should do and can do is determined by the status of the family being served."

In the Harlems of the land there are too few Negro families of "status" to be effective in seeing to it that their young are properly educated. Yet as far as is presently known concerned socioeconomic status is not even remotely significant in determining learning orientation or achievement in the Jewish family and in many other families in the United States.

It is possible, as Ralph Ellison has pointed out, for the Negro to obtain an education if he wants it badly enough. "We possess two cultures—both American—and many aspects of the broader American culture are available to Negroes who possess the curiosity and the taste—if not the money—to cultivate them." Lacking curiosity—money is of small consequence in acquiring an education, as any man of taste knows—no amount of educational reform will bring education to those who have no "taste" for it.

If neither family nor community nurtures a passion for learning, the state can only indoctrinate—whether it be a false historicity and a false sense of values as is now being attempted in certain educationist circles to placate Negro criticism for past errors, or by way of the Little Red Book of Mao. We have not yet reached the latter stage. But the billowing up of waves of newly discovered greatness in a largely mythical Negro African past can be just as unnerving—in the long run.

Dr. Conant also states, "It has been established beyond any reasonable doubt that the community and family background play a large role in determining scholastic aptitude and school achievement." That Negro children respond poorly to testing when compared with white children, is simply a fact. Even a dozen years ago such a statement would elicit the automatic charge of race prejudice. No one was permitted to state the obvious. Today only the liberal who tries to take the sting out of unpleasant truth by circumlocutions refuses to face the fact.

Dr. Alvin Poussaint's belief that "A trained incapacity to be aggressive would also account in large part for Negroes' below-par achievement in school" may be believed or disbelieved. That he recognizes the fact of "below par" achievement is what is important. Once it is acknowledged, perhaps something can be done about it. But hatred and accusations of conspiracy to miseducate Negro children, directed at certain Jewish teachers in New York City, will not help.

Milton Himmelfarb, contributing editor of *Commentary*, is probably right when he says, "In New York today, the educational self-interest of Jews clashes with the educational self-interest of Negroes." Jews, it seems, cannot have enough education, books, art, concerts, and the like. He quotes statistics showing that "a quarter of the buyers of books in the United States are Jews." He concludes that, given enough money, both Negroes and Jews will be able to have education. He really does not believe it though. What he does *mean* he puts into an anonymous Negro's mouth: "You whites have so contrived matters that we Negroes are unable to meet what you are pleased to call your objective, color-blind tests. We will no longer submit to rules that put us to a disadvantage. Admit us to your schools. If that means lowering of standards for you, it will mean a raising of standards for us—and it is the only way for them to be raised." No living Negro ever said precisely that, but that is what many Negroes had in mind when integration was still the main goal.

Early in 1967, Roy Innis, then chairman of the New York chapter of CORE, said, "We seriously doubt there is anyone around who honestly believes that integration is a goal that can be reached in the foreseeable future." Actually, his organization and the other militant participants in the "Negro Revolution," as it has become fashionable to call it, have come to the conclusion that all-Negro schools are the only solution. Mr. Innis thinks that such schools can be as good or better than mixed schools.

Demands for a separate all-black school in Harlem have gone so far that for months the principal of a new $5 million junior high school had to go to his office daily under police escort. "We want black," the pickets shouted. Today Mayor John V. Lindsay, sweating out the problem of education with all the other problems he has, is hopeful that, if he grants the demands of the militant Negroes—he genuinely believes they are the demands of the entire Negro community—for a black principal or two and a few more Negro teachers, civil peace will be maintained.

We have mentioned earlier the charge of "miseducation" directed against teachers who, because they are Jewish and teach in the difficult urban slums, are apt to be rather more dedicated than most teachers in the system. The majority, prevented from maintaining discipline and teaching the subjects that they are paid to teach, must choose between leaving or staying, even though, as I have heard some say: "I don't teach. I can't enforce proper discipline. I am not expected to teach, only to keep order. They (the pupils) don't want to be here, and they hate me because they think I want them here. Why doesn't someone tell the truth?"

What these teachers mean is that many Negro boys and girls would rather be learning a trade or just bumming around or preparing themselves for something that truly strikes their fancy—anywhere but in the classroom, which they regard as a jail. Some of them would gladly serve a three-month sentence rather than spend three years in high school studying subjects that are of no interest to them. Many take out their resentments on the teacher for the simple reason that she (or he) is there. A junior high-school teacher in New York who had his eye knocked out by a recalcitrant pupil was summarily dismissed for failure to keep order. Nothing happened to the student. But society is still hobbled by Depression formulas, which called for keeping young people in school until the unconscionable age of eighteen to help reduce pressure on the labor market. It does reduce such pressure—but at what a price!

Most bright young Negroes can learn more in three months as apprentices in real trades than in three years of vocational high school. But high schools issue diplomas, and diplomas are what the Advertising Council, the Presidential commissions, governors' commissions, and the mayors' commissions say is most essential in life. The word "dropout" has become a dirty word, although some of the best men this country has produced dropped out of school fairly early.

Schooling is for those who want it and put a special value on it. It is not a substitute for work, nor is it or should it be considered as something punitive. Yet that is just how many Negro boys and girls see it. Angry mothers, goaded by racists, shout "We want black only," "You can't educate our kids," and "Black teachers for black children." But they do not understand what they are shouting.

Many teachers now under fire by black racists have only themselves to blame. Some were sincerely convinced that it was right to give generous marks to youngsters for whom they were genuinely sorry. The various teachers' colleges in the land probably brought on this situation. Grounded in almost no discipline, the teachers came to school prepared to indoctrinate the "whole child." But the "whole child" eluded them because he came from a home with values so different from theirs that they never were able to reach him. Instead they played games with the Negro child. They fed him, they dressed him, they sympathized with him, and they pampered him. But they did not teach him. The Negro, however irrational his outburst may seem, is undubitably right about that. As Mr. Ottley sadly observed, "People who start out to do good, often end up by doing bad."

"There is no reason why we should have to join the white community to achieve our goals of education," says Dr. Poussaint. No reason whatever. But the angry black men who would ruin education for elementary-school children by accepting *any* education as long as it is *black* education must not be permitted to have their way. Dr. Poussaint, Dr. Clark, and other Negro intellectual leaders must "tell it like it is" this time. They must reach the Negro mother who knows that her child is behind in reading and arithmetic but can finger-paint to the delight of the local television newscaster. She wants for her child the best education that an affluent society can provide, without frills or substitutes;

she wants her Johnny to read, write, and figure the way other mothers' children can. If her Johnny cannot read, she may as well learn why *now*.

William F. Shockley's charge in February 1967 that "the scientific community has been ignoring or blocking research into possible differences in the genetic makeup of races" should be answered honestly. As *Time* reported the story, "He has been accused, in turn, of fostering racial prejudice." Actually, all Shockley wants is to investigate whether environment alone or in combination with heredity is responsible for the poor showing among Negroes in the Armed Forces Qualifications Tests and others. Dr. Shockley, a Nobel Prize winner in physics, believes that failure of the scientific community to face the problem objectively may be responsible for and contribute to the "decline in the over-all quality of the U.S. population." Shockley attributed uncertainty in dealing with the problem to what he called an "inverted liberalism which, he says, has resulted in taboos against research in genetic differences." He was answered by University of California psychologist David Krech, who said that problems of "measuring racial differences" rather than "taboos" is responsible for the lack of the evidence Dr. Shockley wants. Dr. Krech demands that all comparative statistics take into consideration the difficulty of attaining a comparison in which "all other considerations are equal." Dr. Krech implies that such a situation will never be attained, that all considerations will never *be* equal, but that is not too important. Dr. Shockley is of the opinion that valid scientific criteria can be set up for testing the matter.

Professor Morton Fried of Columbia University would "veto" any such project on the grounds that it "is not a scientific debate at all. . . . There is a need to end practices whereby such studies are treated as serious intellectual endeavors." He was answered by Professor D. J. Ingle, Chairman of the Department of Physiology at The University of Chicago:

> Fried seems to imply that no knowledge which can be misused should be made public. The scientist should be concerned about the use of knowledge, but Fried fails to rationalize the withholding of knowledge that can be used for good as well as evil. . . . It is true that racists seek any possible basis for distortions and generalizations against Negroes which can be used to limit their rights and oppor-

> tunities. This risk is partially averted by emphasis upon the importance of attending to individuality rather than to racial origin, for the range of individual differences within a racial group is far greater than group differences. . . .
>
> Knowledge can be misused; this does not excuse efforts to block inquiry and debate or to deny laymen in a democratic society the right to know. Closed systems of belief can also be misused. . . .

Of course. But Dr. Fried, an anthropologist would, it seems, prefer to close discussion once and for all.

It is of course possible that fear motivates much liberal opposition to the Shockley proposals. Why should the Negro fear it? At worst such tests may prove what many a black activist already believes, that white, Christian, or Judeo-Christian society has no real meaning for the *black man* in the United States, whatever meaning it may still have for the *Negro.* At best they may help to lay a ghost that haunts the conscience of all Americans: Are Negroes really capable of succeeding as a race, as all people of genuine good will hope?

Dr. Conant, for one, argues for further investigation of the matter:

> Until we have a great deal more data about test scores and school records, especially with respect to large numbers of Negro children from stable high-income communities, I for one would reserve judgment on the answer to the question whether there is a correlation between race and scholastic aptitude.

Let us reserve judgment, by all means, but let us not stop investigating out of fear. What are we afraid of? After all, not *all* previous tests have shown a disproportionate difference between Negro and white potentials. In fact, the very first such test, in 1897, yielded startling results; 500 Negro boys responded faster to the test situation than did a similar group of white boys. In the mid-1920s Dr. Carl C. Bingham of Princeton University "proved" with a specially devised test that Jews scored lower than other groups on general intelligence. The results did not cause one single Jew loss of a single night's sleep; not did they stop him from continuing up the ladder of achievement and education, including attendance at Princeton University.

Of course, Dr. Conant is convinced that "Satisfactory education can be provided in an all-Negro school through the expenditure of more money for needed staff and facilities." But such an effort runs counter to the Supreme Court decision; until that humiliating decision is rescinded by a future Court, education in America will be disrupted by brutalities that no amount of "money" or "facilities" can eliminate. The ruling that separate educational facilities are *inherently* unequal is perhaps the most immoral decision ever made in the nation, for it singled out one race for this demeaning assertion. Had the Court ruled, for instance, that an all-Jewish education is *inherently* inferior, it would have been impeached. (The Jensen Report and the studies of Dr. Audrey Shuey do seem to indicate that Negro children have a greater difficulty in the *testing experience for abstract learning* than do the children of all other ethnic groups in the nation. Now this should not mean that these studies should be denied proper discussion or refutation; nor should they be treated as wrong because they are heretical. The social scientist who refuses to air the problem is perhaps more of a "racist" than he would want us to believe, else why the fear. Is it that he has *inner* feelings that opening up this Pandora's box, almost anything may come flying out? If a problem is there it is best to know it—not hide it.)

Some of the bitterness in the so-called "ghettos" has assumed frightening proportions, leading to racism and anti-Semitism in certain New York communities. In Brooklyn Public School 284, teachers became targets of rude anti-Semitic outbursts, their mailboxes were stuffed with crude leaflets, and they were told to leave. The same thing happened at Public School 40 in Queens and at Junior High Schools 250 and 246 in Brooklyn. Teachers were afraid to go to classes, and principals had to be escorted to their offices until the Anti-Defamation League of B'nai B'rith entered a complaint with the Commission of Human Rights, which promised to look into the matter.

Naturally, the teacher wants out. But where is he or she to go? How could he have known that education does not fit everyone as if it were a ready-to-wear garment produced on an assembly line? In the so-called "graduate schools of education" he was taught that every failure to instill learning is the fault of the teacher, not of the pupil.

These graduate schools teach, not only that all people are created equal, but also that all children are *equally* educable. Every teacher who has ever spent a day in front of a classroom knows differently, but who is he to argue with the men who control the licensing of teachers. That Negroes have a grievance cannot be denied. But to correct it we shall have to go back to the situation as it was before 1954 and start over again. If finally large numbers of Negroes choose Stokely Carmichael's solution, which ultimately means education by the Little Red Book, or the McKissick solution, which ultimately means education via the Black Book of Africa, so be it. But let the Negro himself decide, with whatever advice and assistance he asks from both races and in the best interests of all groups that make up the nation.

For proper education—that most important necessity of American youth—a proper family life is needed. As the white middle-class family rapidly disintegrates, it is possible that the Negro family will revive. It will never, however, assert itself as a vital part of the community life until illegitimacy is recognized, as E. Franklin Frazier recognized it, as both morally degrading and despoiling of healthy family patterns. Yet in 1969 we find the figures for illegitimacy growing to a whopping 29.4 in 10 years from a previous high of 23.6. This rise was largely due to the implementation of the Aid to Dependent Children program for families on relief. At the same time New York City's welfare budget had grown to 2 billion dollars—and a third generation of young people on relief were preparing to produce a fourth generation, with the city preparing to set up 20 high schools for pregnant girls—most of them the daughters of mothers on public welfare.

Chapter Fourteen

RACE AND PREJUDICE

It may well be that the South is wreaking revenge on the nation by sending its black citizens north to continue Reconstruction, which some historians have described as "a continuous race riot."

The attribution of revenge to the South, the claim that southerners created impossible conditions for Negro participation in the electoral process because white men had been disfranchised during Reconstruction, may be dismissed, for all revenge theories of history are based on a fundamental fallacy. Group revenge is impossible, and group frustration is inimical to the concept of individual freedom. Prejudice rarely needs "objective" justification. Nor is it wholly or even mainly an affliction of the ignorant. The most educated men have exhibited rancor in race relations and wickedness toward fellow human beings because of race that would defy the capacities of many ordinary men to emulate. It may well be that the so-called "prejudices" of the ordinary man serve as a safety valve (as Edmund Burke suggested) against the crueler "prejudices" of the educated elite.

Most liberal and progressive men hold education *per se* to be so beneficial that they overlook the fact that it does not guarantee decent behavior. It is good to be able to read, but the value of the skill is doubtful if one uses it only to read the collected works of Mao Tse-tung. The Jews, a most sophisticated people in every other aspect of living have suffered more from the myth of the inherent good in learning than have any other people on earth. Some German socialist leaders and intellectuals like Friedrich Engels and August Bebel were even convinced that mere literacy would be a guarantee against the anti-Semitism then taking deep root in Germany.

Only a generation before Hitler hotted up his ovens, Comrade Bebel, in 1906, reassured the Jews in the *Vaterland* and elsewhere that they need no more fear the wrath of the anti-Semites, for the simple reason that in a literate nation like Germany religious persecutions were "impossible." The Jews believed it; they even preached it. When Hitler warned that Jewish heads would roll, the Jews themselves led the chorus of derision. What they had not learned, and perhaps have not learned to this day, is that a literate anti-Semite is worse than one who merely reacts eviscerally when he meets a Jew or someone he thinks is a Jew. There is a parallel among anti-Negro bigots in the United States.

His worst enemies have been not the ordinary men in the streets or even illiterate poor whites. These men have basically responded to theories propagated by the best "progressive" minds of their time. For instance, when the rabid "nigger hater" Pitchfork Ben Tillman invoked the Greek classics to excoriate the "niggahs" it was not important that his audience could not follow his allusions. For example it was not necessary to know the story of Theseus to understand what he meant when he ranted in 1913 that "from forty to a hundred Southern maidens were annually offered as a sacrifice to the African Minotaur, and no Theseus had arisen to rid the land of this terror." In Germany Hitler invoked the Valkyries and the name of Richard Wagner to elicit a similar reaction from a highly literate audience.

In any case, race hatred, group suspicion, xenophobia, and intertribal warfare are as old as time. Man has long shown a propensity to detest what he does not understand and to hate his neighbor, even though it challenges all common sense and all the divine commandments and ethics that wise and holy men have tried to establish. Man "the ornery biped" has continued to suspect nonmembers of his family, group, tribe, nation, or race. I am indebted to Thomas F. Gossett's extremely valuable *Race: The History of an Idea in America* for much of the material in this chapter. A study of race prejudice throughout the ages reveals that almost always the basest of absurdities originated with the most educated men.

In recent centuries the atheists or agnostics have ridiculed belief in the common origin of man in a divinely created Adam by pointing to the "inherent" differences between Negroes and whites, to the disparagement of the former. An early American physician and paleontologist, Dr. Samuel George Morton, undertook, on the basis of his collection of crania from all over the world, the first serious study of race in this country. His work was highly praised by none other than Oliver Wendell Holmes. It was Dr.

Morton's notion that "the key to the separate origin of races was to be found in hybrids and mulattoes." He then went on to "prove" that, although mulattoes are not sterile as are mules, "mulatto women bear children only with difficulty." He then concluded that whites and Negroes are actually different species, not at all related in a common humanity. Many of his researches were confirmed by the distinguished naturalist Louis Agassiz, who also believed in multiple creation.

It was Dr. Agassiz who, in the mid nineteenth century, after observing Negro behavior, predicted that Negroes would ultimately prove the downfall of the Republic. His sentiments paralleled those of Thomas Jefferson, who believed that, unless some way were found to remove the Negro from the land, the nation could not long survive. Dr. Agassiz, however, based his findings on what he considered the strictest scientific research, revealing that the Negro was organically different from the Caucasian.

Dr. Agassiz was opposed in his scientism by the fundamentalist clergyman and slave holder John Bachman of South Carolina, who believed in the monogenic theory of the development of man. But Bachman was no mere preacher; he was an amateur naturalist of some distinction and even collaborated with John James Audubon on a number of occasions. Yet Bachman was correct, and all the scientific humanists, as they were called at the time, were wrong. Bachman, however, went on to justify slavery on the basis of his *theology*, claiming, along with so many other religious apologists for slavery, that Negroes are the descendants of Japheth, cursed to be the "servants of servants." He argued that the Negro is indeed of the human species—but the lowest order imaginable. Bachman's most serious opponent was Dr. Morton, and all distinguished men of science and of "liberal persuasion were on the side of Morton." To them, the Negro was not part of the same species of humanity as, for example, Voltaire, who had held similar views. Some concluded that Negroes could therefore not be enslaved, any more than dogs, cats, or horses could.

Most men of science in those days opted for the polygenic theory, and, as most Abolitionists were themselves "progressives" and tended to accept the more "advanced" thinking of the time, Frederick Douglass found few among them who cared to help him to disprove the racial theories then abounding all over the world. It was enough for them, perhaps, to seek abolition of slavery; to believe that all men were created in the image of one God was too much. To the men of the Enlightenment and to the scientists of the eighteenth and nineteenth centuries belief in God was sheer superstition. They not only refused to believe; they mocked those who did accept such primitive and unenlightened ideas.

The South was adamant in opposing belief in diverse origins. But many northern intellectuals, ever alert to the latest fashions in science and education—became infatuated by the new "science" of phrenology. Men like Ralph Waldo Emerson, Walt Whitman, and Edgar Allan Poe were fascinated by the scientific potential of bumps on the head, as expounded by John G. Spurzheim, whom they hailed as a genius. His theories included the cranial inferiority of certain races. In 1856 Gratiolet in France advanced the theory that in Negroes the coronal suture of the skull closes at an early age, which would account for the "fact" that Negro children can keep up with white children until the "age of 13 or 14," after which they fall behind. This theory is with us still, except that the age of divergence has been lowered to eight or nine and may even account for current infatuation with Head Start programs.

Later, Anders Retzius invented the cephalic index, and the battle of the skulls was on. Retzius' measurements were originally taken on Swedes and Finns, but soon measuring heads, faces, noses, arms, legs, and even genitalia of Negroes—and comparison of these measurements with those of Caucasians—began in earnest; it has not stopped to this day. As late as 1929 Professor Raymond Pearl found that the "temporal lobe of the brain is smaller and differently shaped in Negroes" than in whites. According to Dr. Gossett, "Professor M. F. Ashley-Montagu concludes that the Negro has an average cranial capacity of 1400 centimeters, 50 centimeters less than the white man, but he denies that the difference has any significance as an indicator of comparative inherent intelligence."

After the discovery of the notably small brain of Anatole France (it weighed 300 grams less than the brain of the average Frenchman), the size of the brain as an absolute indicator of intelligence was played down, but its use has not died out altogether. John Fiske, "the eminent historian of science," argued that the size of the brain mattered less than its "convolutions." An infant brain is almost smooth; a thinking adult's brain is full of whorls, wrinkles, and convolutions. A professor's brain is naturally more convoluted, Dr. Fiske argued, than that of a peasant, which is almost unlined. Negroes, he affirmed, have the smoothest brains of all. Dr. Fiske claimed that Charles Darwin had evidence to support this claim. But, as Dr. Gossett says, "Subsequent research has failed to discover any significant racial differences" in these convolutions.

The "science" of today is the "superstition" of tomorrow. Nowhere is this truth more evident than in the matter of racial relations in the United States. To paint a really clear picture of the scientific spirit in the early years of the Republic we must go back to one Samuel Stanhope Smith, ethnologist and minister of the faith in 1787.

Henry Moss, reared in the positivist climate of that time, is worth attention. Moss, a Negro, fought with the Continental armies under George Washington, and lived in the North for a long time, although he originally came from the South. He developed a series of white spots on his body, and in a few years he had become "almost entirely white." He was at first exhibited as a curiosity in Philadelphia, but then the local scientists got hold of him—and discovered the "solution" to the Negro problem. Mr. Smith was convinced that the Moss case was evidence that nurture and climate, rather than genes, determined the color of a man's skin, especially black skin. Dr. Benjamin Rush, a friend of Jefferson and one of the most distinguished men of his time, was also convinced by Moss's case (and independent studies of his own) that the Negro needed only a cold climate to turn completely white and thus to solve the embarrassing problem confronting the humane leaders of a republic in which black men were largely enslaved.

Dr. Rush actually believed that Negro skin color was caused by a mild form of leprosy, which he thought he could cure with common sense, good native medicine, and a cold climate. But Samuel Adams, as if anticipating the present, derided those who, while "yelping loudest" for Negro rights, were often the worst "drivers of Negroes." And Thomas Jefferson, who wrote, in the Declaration of Independence, that all men are created equal, did not believe it himself. "Jefferson," says Dr. Gossett, "will concede only that Negroes are generally kind and humane."

In deference to Mr. Jefferson, it should be remembered that, in his time, only men who believed in God of the divine creation were likely to believe in the equality of all men in the Augustinian sense. Freethinkers, deists, agnostics, and atheists from Voltaire to men of the most recent past have believed only what they could see with their own eyes; and what they could see has apparently not been flattering to the Negro. Voltaire actually ridiculed the whole idea of the equality of races as a dark superstition held only by religious fanatics. But St. Augustine, nevertheless, believed that Jews—even if human—were fit only to be slaves.

So specialized had the science of race differentiation become that Dr. Charles White of England, a leading craniologist, "proved," through the use of graphs, diagrams, and pioneer measurements, that the penis of the Negro is larger than that of the white man. The clitoris of the female Negro was also "proved" to be much larger than that of her white mistress. That is what he believed, and that is what he "proved." He was actually interested in showing similarity between Negro and simian genitalia.

Perhaps it is not so surprising that the scientists who argued the limited abilities of Negroes along racial lines were almost always those who would today be regarded as "progressives." They had rejected God and adopted Reason as the deity of their material culture. The concept of equality of men under the fatherhood of God was indeed alien to them.

Race theorists tried to find support in Social Darwinism which, following Herbert Spencer, preached "the survival of the fittest." But Spencer was of small help because his main belief was that freeing man from responsibility freed him from accomplishing things for himself. (Later certain fanatics of racism found all kinds of comfort in this "survival" theory.) Spencer was opposed to everything free, especially free education and free libraries, on the theory that what one obtains for nothing is worth what one pays for it. William Graham Sumner, whom all progressive sociologists regard as somewhat conservative, echoed this belief, but he was as staunch a supporter of Negro freedom as he was an opponent of all socialist and cooperative enterprise.

Lester Ward, Sumner's progressive opponent, simply refused to believe that Negroes are really equal to whites. He elaborated the notion that Negroes are inferior because they behave as all lower races do, pursuing white women in order to improve their racial stock.

In time the eugenicists, too, got into the race game. Under the leadership of Francis Galton, a cousin of the great Darwin himself, they emphasized the dominance of heredity over environment in genetic development. It was Dr. Galton's opinion that "the average intellectual standard of the Negro race is some two grades below our own." This two-grade difference is apparently recognized by many Negroes today, but they blame it on white teachers who, they believe, consciously try to keep the black man down. And as the twentieth century slowly grew older Tom Watson thundered from Georgia: "What does civilization owe the Negro? Nothing! Nothing!! Nothing!!!"

Dr. Gossett remarks, "In more rational language, eminent leaders in the sciences and social sciences were saying about the same thing."

Apparently no one of stature in this country, except such hard-bitten Yankee conservatives as John Adams and a few others, ever held out much hope for the Negro. Thomas Jefferson would have preferred to be without the Negroes, and Abraham Lincoln despaired of having enough vessels at his disposal to ship them to Panama or any other country in Latin America that would have them. The nation that brought them here against their will must live in the shadow that haunted Benito Cereno in Herman Melville's cruel and perhaps prophetic tale. Theodore Roosevelt invited Booker T. Washington to the White House but later regretted having done so as he came under the influence of the racist writings of Owen Wister. The more scholarly William James was equally influenced, and Henry James, looking at a Negro porter in Grand Central Station, claimed to have understood at a glance the white man's dilemma in the South.

The first man to strip the cant from the "scientific" claims of the professional truth seekers was the self-educated John Mackinnon Robertson, who attracted wide attention with his book *The Saxon and the Celt.* Then came cultural anthropologist Edward Tylor, whose *Primitive Culture* has stood the test of time as have few other works in this field. (Mr. Tylor was not a "professional" either, he too lacked a university degree.) Then came other men who pursued the subject with a minimum of bias. Finally Franz Boas decided to employ his immense energies in the field of cultural anthropology (he had been trained in physics).

But Dr. Boas probed so deeply and was so flexible in his thinking that it is not always easy to determine where he stood on any given day. In one of his later articles, published in *Yale Review* in 1921, he claimed to have solved both the Jewish and the Negro problems. It was Boas' opinion that the Negro problem would disappear when "the Negro blood has been so much diluted that it will no longer be recognized—just as anti-Semitism will not disappear until the last vestige of the Jew as Jew has disappeared." What he meant is that both the Jewish and Negro problems will disappear only when both Jew and Negro as separate identities disappear from the face of the earth. Almost any problem could be solved *that* way.

Apparently none of the scientists discussed here approached the problem of race from a position of conscious prejudgment. All were quite sincere in their investigations.

In any case, it has ever been thus, although that does not, of course, mean that it always has to be thus. The Rig Veda tells us that 5,000 years ago Indra, the god of the Aryans, blew away "with might from the earth and from the heavens the black skins which Indra hates." Thus the dawn of recorded race hatred. These dark-skinned people are described as "flat-nosed," and the Rig Veda goes on to tell how Indra "slew the flat-nosed barbarians." Conquest achieved, he set about flailing the skins of this black people.

Ancient China too had its racial troubles, which we can read about in accounts of the Han Dynasty (third century B.C.). A "yellow-haired and green-eyed barbarian people in a distant province," were a threat to the native population at that time. Egypt's history of racism dates back to the fourteenth century B.C. "When the lighter-skinned Egyptians were dominant they referred to the darker groups as 'the evil race of Ish.' When the darker group ruled, it referred to its adversaries as the pale, degraded, race of Arvad."

Among the ancient Hebrews, Ezra preached the abomination of mixing the seed of Israel with that of the Ammonites and the Moabites. All alien wives and their children were deported, and a strong ban on mixed marriages was instituted. To Professor Ruth Benedict this ban is evidence of "fanatic racism in Israel long before the days of modern racism." A. A. Roback, also an anthropologist, believes that the first recorded slur "against the Negro in the Bible is that of the prophet Jeremiah who asked, 'Can the Ethiopian change his skin or the leopard his spots?' "

The Greeks, who regarded all outsiders as "barbarians," had the legend of Phaëthon, who drove his chariot so close to the earth that it burned him (and all his descendants) charcoal black. Hippocrates, no maker of legends but a physician, attributed racial differences to geography and climate. He was also probably the first to apply the energy theory of ethnic development, claiming that the harder people have to struggle against nature, the hardier and wiser they will become.

In North America the first signs of racial prejudice were observed when John Rolfe decided to marry Pocahontas. The opposition of the English crown was, however, not based on purely racial considerations, as is sometimes assumed. It was based on the curious proposition that Mr. Rolfe, a commoner, had no business trying to marry into royalty; Pocahontas was a princess, even though she lacked a proper kingdom. Then came the Negroes.

It should be clear by now that no one, least of all the scholars of Harvard, Princeton, Yale, and Columbia Universities, has a monopoly on "prejudice." Perhaps the man of common sense is more to be trusted than the man who, in the words of Jonathan Swift, may only be measuring "farts by the ell."

The *refined* language of prejudice is more deeply rooted in the Academy than in the public square. The hefty volumes of anti-Semitic lore are testimony that the finest minds of Christendom, from Geoffrey Chaucer to Feodor Dostoevsky, from the ancient Greeks to the best-educated men of Germany and France, from the leaders of the French Enlightenment to the socialist revolutionaries of Germany and Russia, have joined in abuse of the Jews. But words cannot kill. It is only the fire that the Jews fear. When Negro intellectuals insult them with words, they can laugh it off, but when they are threatened with the "fire next time," the message is clear—and ominous.

part 4

The Confrontation

Chapter Fifteen
THE HISTORY

Why the Jewish-Negro confrontation? Why not the Catholic-, Protestant-, Italian-, Irish-, Polish-, Hungarian-, Swedish-, Japanese-, Chinese-Negro confrontations? The answer is that in addition to the general problem of Negro-white relations in the United States, there is a special relationship of antagonism between Negro and Jew, which neither wanted but from which neither can escape. We need not accept James Baldwin's statement that "The structure of the American Commonwealth has trapped these minorities into perpetual hostility." "Perpetual" is a long time. I hope to suggest improvements in the relations between these volatile minorities from which the whole nation may profit.

I am not aware of any articles dealing with confrontations between Negroes and other minority groups. If they do exist they have certainly not created widespread repercussions comparable to those created by the books, articles, newspaper reports, documents, defenses, and accusations that characterize the Jewish-Negro confrontation. It may be, as James Baldwin has indicated, that "Negroes are anti-Semitic because they are anti-white," but then why are they not specifically anti-Quaker, anti-Swedish, or anti-Catholic? Why "anti-white" necessarily also means anti-Semitic was the topic of an article by Mr. Baldwin in *The New York Times Magazine* of April 9, 1967. Before we take up the Baldwin thesis, however, it is best to review the history of both ethnic groups in the United States, to see how deep the conflict goes.

One need not go back to the earliest historical references to Jews and Negroes for evidence that tension between the "races" has indeed existed for a long time. The earliest literature on the subject is inimical to both; some intellectuals, among them Karl Marx and Sigmund Freud, have even suggested that Jews are of a Negro origin. Marx was not only convinced of the negritude of the Jews; he was equally certain that they were descended from a tribe of lepers that had followed a black Moses out of Egypt into the desert. Freud's study of Moses is evidence that few men excel in more than one discipline at a time. His unfortunate forays into anthropology produced nothing more significant than his curiously uneven work on primitive man, *Totem and Taboo.*

It must be admitted, however, that neither Marx nor Freud helped to foster the mutual suspicion that now intensifies the mutual prejudices of both people. Nor can we rely for an explanation on the work of the Catholic anthropologist Joseph J. Williams, who argued in *Hebrewisms of West Africa* that many American Negroes, especially those in the West Indies, owe their abilities to engage in commerce and trade, as well as other civilized virtues, to the influence of Jews traveling in sub-Saharan Africa in biblical times, especially during the Solomonic period.

Jews have certainly had contact with the sons of Ham and Japheth from earliest times. Today a fairly large colony of 20,000–50,000 black, if not entirely "Negroid" Jews, live in Somalia and Ethiopia. Their acceptance of the Five Books of Moses suggests that, if they are not of Jewish descent, they must have been converted to Judaism at a fairly early period, before the beginnings of the talmudic and rabbinical traditions. Some scholars argue a more recent conversion. I shall have more to say of these Jews later when I take up the issue of color in Israel.

When young American Negroes speak of a glorious black past in Africa, of kingdoms and castles on the banks of the Limpopo River, they may well be referring to some ancient Jewish kingdoms in western Africa. But the "history" that is used to instill the political philosophies of Patrice Lumumba, black nationalism, Mau Mauism, or even an American variety of *apartheid* is not part of the genuine history of Africa. Jesse Woniock, the young American Negro Peace Corps worker in Kenya who, according to *The New York Times* of September 3, 1967, felt that he was regarded with suspicion, despite his color and his efforts to communicate along racial lines, learned to his disappointment that to Kenyan Negroes he was an American, rather than simply a black man. Negro author Richard Wright had discovered the same thing earlier. After the first triumphs of nationhood in Africa, American Negroes, except among some black-nationalist groups, seemed unable to identify with African black men. Nevertheless it has become customary for even so "integrationist-minded" a newspaper as the *Amsterdam News* to identify the Negro as "Afro-American."

However, American Negro identification with Africa seems strangely selective. I know of no organized protest among American Negroes against the bestial atrocities being committed there. I refer, of course, to the Nigerian assault on Biafra, a little country that has sustained casualties numbering in the millions. Can one imagine the Jewish organizations keeping quiet if it came to their attention that thousands—or even dozens—of their fellow Jews were being exterminated? The Carmichaels, the Cleavers, and all the others who seek artificial identification with Africa are "cop-outs" from responsibilty; they attempt to delude masses of Negroes into a concern for those African states that are least independent, *least* viable, and most totalitarian.

The anti-Israel venom of both Islam and international Communism—whether Maoist, Brezhnevist, Stalinist, or Khrushchevist—provides a model for anti-Semitism in the United States. President Felix Houphouet-Boigny of the Ivory Coast who hailed the contribution of "Hebrew thought and culture in the development of a civilization which the Ivory Coast is happy and proud to share," would probably be as shocked by H. Rap Brown as would any cultivated man in the United States—maybe more shocked. It has become customary for American white men to bend over backward to accommodate scurrilous attacks from Negro extremists.

Neither ancient nor recent troubles of Israel are important to the study of the Jewish-Negro confrontation in the United States. Neither are the facts that the first twenty-three Jews to arrive in New Amsterdam, in 1654, were not all rich or slumlords, and that the first Negroes to arrive on these shores were not slaves. What is significant is that, from small beginnings the Jews went on to become rich and powerful citizens of their adopted land while the Negroes were soon turned into slaves and have never been able to reach a degree of integration into our society comparable to that of the Jews. But the story is not yet finished.

That each people has its own traditions different from others' traditions is fundamental to our thesis. People with different life styles will respond to altered circumstances in different ways. Only a totalitarian, a man who believes that his way is best for all, even at the price of extinction, cannot grasp this point. The fact that the victims of one of the first two lynchings in the South (in 1868) were a Negro and Jew, who were hanged on a common gibbet, has not increased mutual tolerance; nor has the Jews' central role in the civil-rights struggle immunized them to the hatred and distrust of the Negro community.

Prejudice is a two-way street, and people find it hard enough to live with neighbors of similar culture, let alone with those of different race. Flemish-speaking and French-speaking Belgians despise each more than each hates the common enemy at any given historical moment. The same is true of French-speaking and English-speaking Canadians. Hausa despises Ibo in Nigeria, as brown-skinned Bantu despises all black Africans. Emperor Haile Selassie of Ethiopia may speak of an Asian-African bloc for political and state purposes, but his royal family regards itself as pure white and Caucasians as merely pink. According to an Ethiopian legend, God originally contrived a clay mold in which He cooked the first man, who came out half-baked, the white man. Then he roasted another man and overcooked him, and he came out charcoal-black, the Negro. Finally, He put in the last man and braised him at the proper temperature with the right amount of seasoning; he emerged neither too black nor too white but exactly right, the Ethiopian. And so it goes.

Prejudice does not die simply because its irrationality can so easily be demonstrated. All attempts to achieve one world, the brotherhood of man, the classless society have so far failed. The leveling effects of education have turned out to be chimerical. Yet the more the world turns to utopian ideologies that promise freedom and equality for all, the more it divides into nation-states, classes, races. Young Negroes openly proclaim black to be superior to white; the apostles of the "new politics," with young Jews in the vanguard, rush to agree in an example of reverse Uncle Tomism the like of which the nation has never seen before. That these young Jews do not mind indicting Israel "as the imperialist aggressor" in its war with the Arab states is not at all surprising. But that they should shout approval for a resolution calling for black liberation by "any means possible" is troubling, for such liberation could lead to their destruction, whatever happens to non-Jewish white liberals.

What is significant, as in so many other areas where prejudice is programed, is that the intellectuals have rationalized acts of violence. From this approach the Jew outside Israel usually suffers most. Before the turn of the century Russian Jewish boys left the *heder* (the Jewish elementary school) to join with their Russian comrades in the Party of National Liberation of the People's Will (Narodnya Volya); and passed out anarchist-inspired leaflets calling on the workers and peasants to "kill the Tsar, the feudal Lords and the Jews." Their role did not save their fathers and mothers from the pogroms inspired by those leaflets, but history fails to record the wanton killing of more than a single tsar or a few lords of the manor.

The *heder* boys who lived to see the revolution triumph, soon found *their* language, *their* religion, and *their* people banned as agents of imperialism. Today young American Jews are helping to bring on massive pogroms against the Jews of America—a process which has been detected by some as going on for a number of years but which has not yet found the official or semi-official sanction of accepted attitudes by large numbers of people, black and white. But that too is a danger.

In any case, early American records fail to show any evidence of Negro hatred of Jews, although Jewish attitudes on slavery tended to follow regional prejudices and beliefs. The earliest surviving writing on the Jews and the Negroes is a strange little volume by Arthur T. Aberneithy, *The Autobiography of a Mad Man.* Mr. Aberneithy argues that the Jews are actually Negroes in disguise, for, like Negroes, they whore after strange gods and strange women. As proof he quotes from Numbers 12:1, "And Miriam and Aaron spoke against Moses because of the Ethiopian woman whom he had married; for he had married an Ethiopian woman."

A more direct exposition of the relationship between Negroes and Jews did not appear until 1904, when Professor M. J. McGhee, director of anthropology for the St. Louis Exposition expressed the following sentiments: "The Jew today is essentially a Negro in belief, physical peculiarities and tendencies. . . . Like the Negroes, they have foisted race riots . . . and in temporary ascendancy have manifested their control by austerity and criminal brutalities." Whatever the professor meant, he summed up with the hope that "after the Negro is eliminated from citizenship, the American government will turn its attention to the alien Hebrew. . . ."

These remarks aroused no protest—then; it was apparently taken for granted that certain ethnic specialists had license to express their prejudices. The Jews, nevertheless, continued about the business of acquiring education and affluence, and the Negroes, denied the franchise in the South, voted with their feet and came flooding into St. Louis, Chicago, and New York in numbers far exceeding anything post-Reconstruction southerners had expected. Besides, neither Jews nor Negroes cared much what the intellectuals were saying. They probably cared not at all for the formidable defense put forth by Franz Boas in their behalf, a defense of Negroes' rights to be non-Negroes and of Jews' rights to be non-Jews.

In any case, the first faint tremors of the forthcoming confrontation between Jews and Negroes were detectable in the words of no less a figure than Feodor Dostoevsky. Suddenly, and probably for the first time in history, the Jew is charged with exploiting the Negro in the same language in which he had once been reviled for usurious practices by white Christians. From Dostoevsky's *Diary of a Writer* we learn that he had long wondered what would happen to the poor and simple Negroes exposed to Jewish perfidy in the United States after the Emancipation Proclamation of Abraham Lincoln. Fortunately, he reminds us, he had not long to wait. A dispatch in *Vestnik Europi* had soon told him the worst: "American Jews in the Southern States have fallen *en masse* upon the millions of liberated Negro slaves" and were preparing to enslave them anew in their own golden chains.

Before Dostoevsky, most allusions to Jews and Negroes were invoked for invidious comparison. But he introduced views that helped to bolster *anti-Jewish* prejudices. (In our own time the theme has been repeated in a series of anti-Semitic articles in *Liberator*, an organ of black nationalism in America.)

Otto Fenichel, in a psychoanalytic study of anti-Semitism has pointed out the need to beware of the "rational" motivations behind anti-Semitic assertions. (The Jew is hated because he is rich, despised because he is poor; he is always ritually washing himself; he is dirty; he is too smart; he is too stupid; he has no creative ability; he monopolizes art; he is a do-gooder; he is an exploiter; and so on.) "The anti-Semite arrives at his hate of the Jew," says Fenichel, "by a process of displacement, stimulated from without. He sees in the Jew everything which brings him misery—not only his social oppressor but also his own unconscious instincts, which have gained a bloody, dirty, dreadful character from their socially induced repressions."

One may deduce that, although Jewish boys and girls were in the vanguard of those "going to Mississippi," the Jew is now in the unpleasant and dangerous position of scapegoat for the Negro. As before, there will be no one to protect him.

We need not be surprised therefore at the remarks of CORE leader Clifford H. Brown, who, at a public meeting in Mount Vernon, New York, on February 3, 1966, blurted out that "Hitler made one mistake when he didn't kill enough of you Jews." James Farmer's mild rebuff to his then subordinate prompted Will Maslow of the American Jewish Congress to resign from CORE, stating in the *Congress Bi-Weekly* that "racism is racism."

This episode came shortly after the Watts riots, in which many Jewish-owned stores were sacked. It was then that the *Jewish Daily Forward*, the largest Yiddish-language newspaper in the United States, referred to the riots as "pogroms." After the ashes had cooled in Watts, a young Negro pointed out to a television reporter the "Jew store" that he had helped to pillage and burn. A year earlier the leader of the Philadelphia N.A.A.C.P. had denounced all Jewish leaders in the civil-rights movement as "a bunch of phonies." Challenged to modify his remarks, he asserted he meant "all so-called Northern white liberals." He went on, however, to berate the "thieving merchants" as the worst of the Negroes' exploiters and added: "The only Negro store (in the Philadelphia riots of 1965) was owned by a man named Richberg. They thought he was a Jew."

One might conclude that the Negro indictment of Jews as "liberals" and "thieving merchants" is contradictory. But it is this very dual identification of the Jew that is at the root of much open and concealed hatred directed at him. Hoping to transfer a compassion denied him when he was the underdog, he can be found in the forefront of the struggle for rights—any rights—as long as they can be identified as "progressive." He is also known for the property he owns but does not live in; for his grocery in a neighborhood that is generally hostile to him for his financial success is more often than not regarded with suspicion by the less successful; finally, he no more wants "racial balance" in his own neighborhood and in his children's schools than does the rest of the white population.

Despite the efforts of certain well-meaning Jewish spokesmen, the Jews in this country may be in for a very hard time just because the Negro sees them both as exploiters and as benefactors. This dual identification was visible in Germany before the holocaust, when the Jew was denounced as both Communist and capitalist in the same breath. To the Negro in the American city today the Jew is as visible as is the Negro to the white community as a whole.

Chapter Sixteen

HARLEM

When Langston Hughes looked about him in the Harlem of the 1920s, he saw the Jewish pawnbroker and old-clothes men, and he composed a minor blues, called "Hard Luck," which counseled those overtaken by hard luck to sort out their finest clothing "An' sell 'em to the Jew." Actually, Hughes' poem is no more insidious than the Yiddish refrain that tells us that the Gentile is always tipsy (*Shicker is a goy*).

A generation later James Baldwin wrote: "Just as society must have a scapegoat, so hatred must have a symbol. Georgia has the Negro. Harlem has the Jew." What Mr. Baldwin said is true neither for Georgia nor for Harlem. He poses as *philosophe* of the symbolism of hatred. And the philosophe—in Paris, in Harlem, in Greenwich Village—has always been a menace to those truths that are denied by the symbolism of hatred.

In the Harlem of Langston Hughes life was rich, and it represented the Negro's New Jerusalem. When Claude McKay came "home to Harlem," he wrote, "Far and high over all, the sky was a grand blue benediction, and beneath it the wonderful air of New York tasted like fine dry champagne." The "wonderful air" does not taste like "champagne" anymore, but the Harlem he rediscovered was no "ghetto," and it was no slum. James Weldon Johnson, the singer of black man's songs, thought it the most delightful area in all New York. During Prohibition Harlem was Saturday night for many of us who went there in the 1930s. We clubbed on Lenox Avenue, yet nowhere did we hear the terms "black," "white," "Caucasian," "Afro-American," except in those small and out-of-the-way places where the cranks gathered. There were, of course, the foot-weary comrades from downtown, trying desperately to wean the Negroes from Harlem into a community of hatred and separatism, by preaching a doctrine of "self-determination in the Black Belt." When some young

Negroes wanted to attack them, those of us with a sense of humor were able to avert trouble by the simple expedient of telling the Stalinist fanatics to depart in haste.

I cannot now recall just who was black and who was white in our club on Lenox Avenue. It simply did not matter; we preferred to have fun, to argue, and to enjoy the best food in town. When the first rumblings of Negro discontent began to be heard, in pious pieces by local writers who asked Negroes to emulate the Jews, a hostile quality entered our discussions. Soon the war and the new intellectualism intervened to dry up our delight in an area that was still one of the finest places to reside in the great metropolis.

The Negroes in New York did not always live in Harlem. Harlem, that is, Central Harlem was special territory. In fact, the Negroes who lived around City Hall, Greenwich Village, or the Tenderloin in the West Fifties regarded it as a great triumph when the first thirty-five Negro families moved to 135th Street. The well-to-do Jews who lived in the area (the poorer Jews lived in East Harlem, where the Puerto Rican *barrio* now is) warned that the first thirty-five would not be the last and that soon Harlem would be all Negro. Their fears were "shown" to be unrealistic by the leading sociologists of the time, but the professors proved to be wrong; within a short time Harlem had become the Mecca for Negroes coming North.

During the 1920s, the time of the Great Experiment (as Prohibition was called), Harlem was the center for the arts in the city. Negro poets, journalists, composers, artists, and actors abounded; it became fashionable to cultivate a kind of spurious Negroism in the Carl Van Vechten manner. African sculpture was "discovered" at that time, and the American Negro was immediately associated with its creation in the minds of white intellectuals. Negroes who had no particular affinity with Benin bronzes were given credit for a culture that was as alien to them as it was to Mr. Van Vechten. All the while the Cotton Club flourished, the Apollo played to standing room only, and no place was gayer than Harlem, U.S.A.

Chapter Seventeen

THE AMERICAN JEWISH COMMITTEE REPORTS

World War II brought many more Negroes to the factories in the North, and soon a kind of congestion of the spirit set in. Right after the war a dialogue was opened, very likely at the suggestion of the American Jewish Committee (A.J.C.), which was sensitive to rumblings of Negro protest at what they regarded as "Jewish practices."

Jewish housewives in the Bronx had been accused of reintroducing slavery or peonage in their treatment of Negro household help. (I heard this charge repeated on a radio program with New York's Commissioner of Human Rights in 1966.) There had been riots in Harlem and in Detroit, not serious perhaps but cause for concern at the rapidly deteriorating relations between the two minority groups. In February 1946 the brilliant young Negro psychologist Kenneth Clark published in *Commentary*, the organ of the A.J.C., an article entitled "Candor About Negro-Jewish Relations," outlining four reasons for anti-Jewish feelings among Negroes:

1. Anti-Semitism offers a pretext for release of aggressions caused by insecurity and humiliating status.
2. It serves as a vehicle for antiwhite feelings in general; often the term "dirty Jew" means "dirty white."
3. It permits feelings of solidarity and identification with the dominant white group against another minority.
4. It helps to strengthen deflated racial self-respect; for example, brazen anti-Semitism is an integral part of the appeal of some intensely chauvinistic Negro organizations.

In 1958 the A.J.C. included this portion of Dr. Clark's article in a sort of confidential report entitled *Negro-Jewish Tensions*, without challenging his simplistic formulations and crude apologetics. Certainly the A.J.C. might have pointed out that the Negro might equally substitute "dirty wop" or "dirty mick" for "dirty white," insulting groups that had never gone out of their way to help him?

The report continues, "Hatred of the Jew, according to the *perceptive* James Baldwin, is the best form the Negro has for tabulating his long record of grievances against his native land" (emphasis mine). That the authors of the report should call Mr. Baldwin "perceptive" on the basis of this remark can be accounted for only by their own probable acceptance of it. They also comment that Richard Wright "hated Jews," not because he was exploited by them, but because he had been so indoctrinated in the "Christian church."

The quotations suggest that both Mr. Baldwin and Mr. Wright (and LeRoi Jones at a later date) are projecting their own private hatreds of the Jews, who, more than any single group in the United States, helped them to begin their literary careers. The apparent paradox reflects the ambivalence of Negroes who cannot accept as disinterested Jewish solicitude for the oppressed members of a race not their own.

Ralph Ellison expressed mild contempt for Irving Howe's concern at Negro misery. In an exchange of letters in the *New Leader*, the distinguished Negro writer told how "uncomfortable" he feels whenever certain Jewish intellectuals write "as though *they* were guilty of enslaving my grandparents, or as though the *Jews* were responsible for the system of segregation." Mr. Ellison was demurring from Professor Howe's notion of the "clenched fist" attitude that he believes the Negro writer must assume if he is to fulfill himself.

The same paradoxical confrontation is apparent in the debate, published in *Commentary*, between Professor Paul Danzis of Columbia University, upholding black power, and Bayard Rustin of the A. Philip Randolph Institute, arguing against it. But on a less exalted level the Negro *Los Angeles Herald-Dispatch* editorialized a number of years back: "Each time the black man tries to establish his own leadership he is blocked by the Jews who fear an ultimate economic loss if the Negro takes over his own destiny. Our main task . . . is to rid ourselves of this phony Jewish leadership." The specific complaint was the almost all-Jewish leadership of the N.A.A.C.P. Yet the Jewish leaders of the N.A.A.C.P. refused to heed the warning—perhaps "appeal" is the correct word—much as Irving Howe later refused to heed the arguments of Mr. Ellison. By 1967 the Negro novelist was reiterating them at the top of his lungs, almost as if goaded into stridency by the deafness of Jewish intellectuals.

Dr. Kenneth Clark actually assessed the problem better than did the A.J.C. Committeemen who issued the report. In the 1946 article already discussed, Dr. Clark says:

> More Jews have shown active concern about racial problems and more Jews have been willing to hire Negroes for various types of jobs. The Negro's interpretation of this has not, however, been altogether favorable. There is sometimes the lurking suspicion that all this is motivated by a desire on the part of the Jew to use him as a shield and reflects a not too well disguised concern about his own status.

A more forcible statement was made by Horace R. Cayton, co-author (with St. Clair Drake) of *Black Metropolis*, a book much quoted by all interested in Jewish-Negro relations: "There is . . . resentment on the part of the Negro that the Jew is so often willing to fight the Negro's battles, but is often reluctant to fight his own." This statement is so true that the A.J.C.'s failure to understand it is astonishing.

The report is candid enough to comment that Professor Clark also objects to what he considers condescension in the Jewish concern for Negro rights. Referring to an institute of Judaism and race relations held in 1945 by the Central Conference of American Rabbis, he wrote:

> On the face of it, it appears commendable that one minority group should be concerned with the status of another oppressed group. But the question arises as to what Jews and others would think if a conference of Negro leaders were to devote a round table to the problem of The Jews in the United States.

Yet rabbinical and secular conferences are still taking place, for certain Jewish intellectuals, in their arrogant concern with other people's suffering seem to court humiliation and self-destruction. No matter how many times the Negro says "go away" the Jew insists on his mission. The American Jewish Congress put out an apologetic little pamphlet entitled "A Reply to Certain Slanders," which pretends outrage that S.N.C.C. should have singled out Israel from among all nations for its contempt. In a foreword, Congress President Rabbi Arthur J. Lelyveld insists that he will always remain a partisan "of all those who seek full equality"; he cannot suppress his indignation at an organization that he had been "proud to join in its campaign of non-violence," and he adds that he will be "forever grateful that the movement which S.N.C.C. was then a part" of, afforded him opportunities to serve the cause.

Yet after Mount Vernon, the S.N.C.C. betrayal, and a host of similar experiences he still asserts, as does his pamphlet, that "no one will separate the Jewish community from its brothers in the great American struggle for racial justice." Of course, Rabbi Lelyveld believes that he and the Congress speak for American Jewry, whereas they may represent only about 1 percent of it. It seems that nothing Kenneth Clark, James Baldwin, James Meredith, Eldridge Cleaver, Floyd McKissick, or any other black American can say will dissuade certain Jewish leaders from intruding into an area where they are unwelcome.

The A.J.C. report expresses its bewilderment that "A Negro professional active in communal affairs in Harlem complained privately that Jews were 'paternalistic' toward Negroes. Jews, this Negro alleged, by taking the initiative in programs to help Negroes *were actually depriving them of the chance to develop their own leadership*" (emphasis mine).

This charge was a serious one in 1958, and it is a serious one today. It is possible to draw up a bill of particulars indicting certain Jewish organizations and intellectuals for helping to exacerbate relations with the Negro community to such an extent as to make any future dialogue almost impossible. Dr. Alvin Poussaint and Mr. Rustin believe that young Negroes have actually been cheated of their chance to develop by the young Jewish intellectuals who came to the South and "took over," to the dismay and embarrassment of the untutored Negroes who were still groping for leadership.

Yet the Synagogue Council of America declares: "As Jews, we are committed by our faith to work for racial justice in an integrated society. Any Jew who fails to join in this struggle demeans his faith." I do not know what faith the Synogogue Council means, but it cannot be the faith of Jews who believe in God, themselves, and their country. I marvel all the more that this statement is quoted approvingly by Charles E. Silberman, a man who is usually little given to cant. Even more startling is Mr. Silberman's notion of holy writ. Describing his horror at hearing Clifford Brown's statement about Hitler and the Jews, he claims that the following night he went to *shul*, heard a portion of the twenty-third chapter of Exodus, and suddenly understood things better. The portion reads: "If you meet your enemy's ox or his ass going astray, you shall bring it back to him. If you see the ass of one who hates you lying under his burden . . . help him to lift it up." Mr. Silberman may not be aware that the Lord's admonition was to help the *dumb* creature. The scripture does not suggest reprisals on the dumb ox or ass of an enemy, because it cannot help itself. Negroes are not, however, oxen or asses. They *can* help themselves. If they ask for assistance, it should be generously given. The Bible also admonishes that "If a man shall steal an ox, or a sheep he shall restore five oxen for an ox, and four sheep for a sheep." The meaning is clear. When it comes to man made in God's image he must be responsible for his conduct.

It is mostly where Jewish contact with Negroes has been closest that anti-Semitism is most noticeable. The A.J.C. report notes: "The tension in Negro-Jewish relations was mostly one-sided; the Negroes resented the Jews. . . . In sum, everything Jews do is apt to evoke Negro criticism or suspicion."

James Baldwin repeated the message for the umpteenth time as recently as April 9, 1967, when he wrote: "The Jew does not realize that the fact that he has been despised and slaughtered does not increase the Negro's understanding. It increases the Negro's rage." "(The [Jewish] liberal) is our affliction," Mr. Baldwin told a group of Jewish intellectuals, as reported in *Commentary* of March 1964. Yet even when Negro intellectuals pronounced, "The civil rights movement as we have known it is dead," only the Jewish intellectuals refused to go to the funeral.

Chapter Eighteen
PATERNALISM AND PATRONAGE

In the running dialogue which Jewish and Negro intellectuals have been carrying on for a generation only Charles Silberman has dared to recognize the self-abasement of the white liberal, particularly the Jewish liberal, in his confrontation with his Negro counterpart. In a round-robin discussion he said that "the white liberal's greatest burden and greatest handicap (is) his almost pathological—and certainly pathetic—desire to be loved by Negroes." He concluded with these remarks:

> The need to be accepted prevents the white liberal from speaking his mind, from disagreeing with Negroes. But of course there can be no real or honest relationship so long as whites fear to speak their minds, or so long as they say one thing in private and express another in public, because they don't want to disagree with the civil rights leaders. This refusal to speak honestly, this refusal to disagree, this persistent concern for "maintaining our relationship" seems to me to be in many ways the worst form of patronization of all.

He wrote these words in 1966. Today we are reaping the fruits of this "patronization," of the distrust engendered by dishonest silence, of the cowardice so often exhibited by liberals. We have reached an impasse in which persuasion has become impossible. Today almost all of yesterday's civil-rights moderates speak the harsh language of power and racial distrust, from fear of losing their slight hold on certain young Negroes.

The permissiveness of liberal Jewish social workers has often led to complete breakdown in morale. The child who is not immediately rebuked when he is rude will become ruder and more insolent as he grows older, as every educator knows—or ought to know. But social workers seem never to have heard that appropriate rebuke is something they *owe* their client as their parents owed it to *them.* The following passage is quoted from the 1958 American Jewish Committee report:

> When Jews moved from the area (East Bronx), the religious school ceased to function, but the institution continued its other services to the Negro community (which had replaced it). The director of Council House, testifying in hearings before the New York Metropolitan Council on Fair Employment Practices in 1943, said that the National Council of Women "has always been very liberal" and that was why it would continue serving the community. Some of her testimony is illustrative of the attitude of Jews and Negroes toward each other. Speaking of the change in clientele from Jewish to Negro children, the director said:
>
> "Our problems are a little bit different. . . . I found they weren't easy to handle. I don't know if this is true generally, but I find their interest span is shorter than the children we were accustomed to dealing with. We find they are untruthful. They have no compunction about lying and taking things. . . . It seems the general ethical concept of the child is a little lower than what we are accustomed to dealing with."

Presumably the director eventually made it clear that all children benefiting from the institution would have to comply with certain minimum standards of conduct. Otherwise her report might never have been made. But had she cracked down on the first child who lied to her, who stole, or who was rude—(she tells later how she was accosted by a Negro boy who called her a "white bitch," which she dismissed with a tolerant smile), she might have been greeted with respect instead of hostile contempt. The boy was really asking for his *human right to be punished.* Asked why she put up with it, she replied, "These are the things you just take as a part of being a settlement house worker. . . ." But one does *not* take them from everybody—only from those considered "inferior" little savages it is one's duty to "uplift."

The A.J.C. summed up its optimistic findings in these words: "Barring a serious depression, it is probable that the economic and political status of the Negroes will continue to improve."

The serious depression has not materialized, but much Negro violence and discontent have, despite (or perhaps because of) economic and political improvement. The kindly Jewish gentlemen who chose not to depart in peace are reaping a whirlwind of hatred. Now when certain Negroes attack Israel and Zionism, leftist Jewish intellectuals tell them that it is possible to be anti-Israel or anti-Zionist without at the same time being anti-Semitic. This message has been preached by the world socialist movement from its inception. That this anti-Zionism is often a euphemism for anti-Semitism (as it is in Russia and the Soviet satellites and in certain Arab circles) is also true.

Daniel H. Watts, editor of *Liberator,* attended the Newark Black Power Conference in 1967. And in that journal Eddie Ellis' articles on "Semitism" in the black ghetto appeared. These articles, with their appeal to hatred of Jews and their attacks on Jewish philanthropy, prompted Ossie Davis and James Baldwin to resign from the advisory board of the publication. LeRoi Jones was at the conference too. Mr. Jones is a prime example of what Jewish tolerance ("exploitation," the black militant would call it) can achieve in a fairly short time.

Only a few years back he was teaching at the New School for Social Research, writing verses, and editing small anthologies of poetry in which few Negro writers were represented. Then one day he appeared in the columns of *Midstream,* a publication sponsored by the Herzl Institute of the official Zionist organization in the United States. The (ZOA) editor thought it fitting to acquaint his readers with the Black Muslim movement. Mr. Jones wrote on American foreign and domestic policy, and his approach was so far to the left that I warned the editor then that his new contributor would soon be in the forefront of those Negroes who find it terribly rewarding to attack Jewish liberalism. We had not long to wait. Soon Mr. Jones published a number of poems in *Liberator* and several plays so anti-white and so arrogantly anti-Jewish that even Jewish liberals were moved to protest. His poem "Black Art" celebrates:

> . . . . *Another bad poem cracking*
> *steel knuckles in a jewlady's mouth.*

The poem delivered its message with a bang, and Jews wondered what they had done to make Mr. Jones so angry. But Tom Griering, a contributor to *Liberator*, put the matter in proper perspective when he commented on Mr. Baldwin's resignation from *Liberator:* "Obviously whitey, being a classic sadist, is also, though more secretly, a masochist. He likes to be told how disgusting he is all the time; he wants to be whipped."

We have come across this phrase before, in a report from Mississippi by Pete Hamill. It appears that some Jewish intellectuals have a special penchant for that kind of self-flagellation. In one issue of *Liberator* (May 19, 1967) Umoja Kwanguvu wrote: "The black man does not live by Levy bread alone; therefore, you don't have to be Jewish to exploit Black people—but it does help. . . . I say to the Jews who are angry about being singled out that, if the exploiter yamulka fits, wear it gladly and be thankful you don't need helmets." This epistle to peace and brotherhood appeared in the same columns with a letter from a Jew, in which he apologized for *his* white skin and assured the editors of his total support for the Black Revolution.

At a "speakout" in Greenwich Village the martyrdom of Andrew Goodman and Michael Schwerner was dismissed on the grounds that they went south to "assuage their consciences." When one of those present rose to protest mildly and made passing reference to the 6 million Jews killed by the Nazis, Archie Shepp, jazz musician and one of the principal speakers at the meeting, retorted that he was "sick of you cats talking about the 6 million Jews." LeRoi Jones told the audience that "our enemies" include "most of you who are listening." According to *The New York Times* of February 11, 1965, most of those present were white liberals. Very likely they were also mostly Jewish.

At long last the editor who had first opened his Zionist columns to Mr. Jones felt constrained to write, "What is painful to note is that there also exists a handful of young Jewish neo-leftists, who joined on the platform in the course of his ghoulish exhumation of the memory of the civil rights martyrs for purposes of ridicule."

What is perhaps even more painful to note is that the editor helped to make a *mentsch* out of LeRoi Jones by publishing him and other black rabble-rousers in the pages of his journal, which is completely subsidized by Zionist funds.

I have been told that James Baldwin never had his current notions about Jews and liberals until the liberal editors of *Commentary*, which is supported by the A.J.C., opened their columns to him. When Baldwin finished one of his early pieces for that secular Jewish publication with his remark, quoted earlier, that the Negro looks at the Jew in Harlem much as the white man looks at the Negro in Georgia, Nathan Glazer wondered could that be *all* that Jewish effort has meant to the Negro?

Even Ralph Ellison has now been moved to outlandish accusations against the editors and contributors of *Commentary*, calling them, of all things, "segregationists." And the Negro *Amsterdam News* dispatched a reporter to Israel to seek examples of discrimination against its darker citizens.

Chapter Nineteen

ORIENTAL JEWS: THE NEGRO AND ISRAEL

Early in 1967 James Booker summed up his two-part series on Israel in the *Amsterdam News* by observing that "often in traveling and meeting leaders of the Ashkenazi (East European) and Oriental groups here we heard references to two Israels." (The word "oriental" is used here and in Israel to mean any Jew not of European or American origin.) The reporter quotes an unidentified "oriental" leader to the effect that "Israel will never be a country if it is ruled by Western Jews." The meaning is obscure, but the reporter's intentions are not. *He* supports an oriental Israel, probably because he thinks that the darker-skinned North African, Yemenite, and Indian Jews will be more favorably disposed to the surrounding Arab states, which is about as far from Israeli reality as it is possible to venture.

According to my own firsthand investigation conducted during the spring of 1967, the so-called "oriental" Jews are more inclined to intransigence on this score than are the Israeli leftists, who are all (or were before the six-day war in June 1967) pro-Arab. Of course, when Mr. Booker refers to the "European *and American*-born ruling forces" (emphasis mine) in Israel, one is compelled to ask whom he has in mind? Except for Golda Meier (who came to Israel from Wisconsin), there are very few Americans in high circles of the Israeli government. Most American Zionists do not want to settle in Israel, and former Premier David Ben-Gurion has chided them in rather harsh terms. A witticism current in the American Jewish community defines a Zionist as a Jew who raises money from another Jew so that a third Jew may go to Israel.

Mr. Booker is also of the opinion that dark-skinned Jews are discriminated against in what he describes (quoting an unidentified source) as "an ugly manifestation of racial prejudice and the Ashkenazi Jews, in an ironic example of cultural interchange, have adopted the racial attitudes of their European persecutors." He cites some statistics (whose source is not identified) showing that "12 percent of the Israeli Jews are illiterate." It is likely that almost the whole "12 percent" is female and that literacy was not considered desirable for women in the communities from which they migrated.

Mr. Booker concluded that, "while Israel's leaders contend they do not have race prejudice, there is definite class and culture prejudice. In many parts of the nation, the oriental Jews live in almost segregated areas and there is little communication among the older population."

What Mr. Booker means by "class and culture prejudice" quite eludes me. As the regime is largely socialist, at whom is the "class prejudice" directed? The reporter is on more solid ground when he speaks of "culture prejudice," which is a commonplace of life the world over and not necessarily to be deplored as in our innocence we once deplored the differentiation of the American population.

Most Israelis appear to be smitten with the "cultural" traits of their less Westernized coreligionists—from the appetite for *falafel* to belly-dancing in Beth Sheba. Perhaps the greatest prejudice I found there was that existing between two communities of Ashkenazi Jews: the ultra-orthodox who live in the Mea Shearim section of Jerusalem and the *sabras*, whose contempt for the bearded, side-curled, and behatted Jews is genuine if somewhat misguided. This prejudice was real and visible and it has nothing to do with race or origins. It has everything to do with religion, which is a far more formidable problem for Israel than is color prejudice.

As for "segregation," Mr. Booker fails to grasp the fact that people like the Jews in Mea Shearim want to live by themselves, among themselves, and for themselves. As long as they do not interfere with the rights of others (unfortunately not always the case), they have every moral right to do so.

It is true, however, that the rate of unemployment among darker-skinned Jews and Arabs is considerably above the national average. Although such Jews constitute more than 50 percent of the total population they hold fewer than 10 percent of the civil service jobs. But, in view of their almost total lack of administrative, technical, and clerical skills, what could they do in the civil service? There is really no purpose in putting them on the already swollen "Federal" payroll.

Mr. Booker concludes: "*In many ways we found their problems similar to those faced by Negroes in many parts of the United States.*" He is quite mistaken. He is, of course, talking about the darkest-skinned (though only occasionally Negro) segments of the Israeli population. Their problems are not at all similar to those of the Negroes in the United States. There is indeed suspicion and mutual antagonism among Jews, but they have always been part of the folkways in exile or at home. East European Jews hold mutual antagonisms for one another that startle outsiders, who can never understand the deference with which a *Litvack* (a Lithuanian Jew) speaks of the Polish Jew and the contempt both hold for the *Galitsianer*, or Austrian, Jew. The contempt is multiplied tenfold when all three speak of the *Yecki*, or German Jew. The manner in which a Texas blowhard speaks of a resident of almost any other state and the contempt of the Yankee for the rest of the population are comparable but much milder versions of the same phenomenon. Such intolerance is in the nature of man, regardless of whether or not we like it, but it is still a far cry from the kind of separation of the races that Mr. Booker seems to believe is part of the Israeli reality.

Not that the mother of a nice pink-and-white Jewish girl in Tel Aviv would look with favor on her daughter's marrying a dark Yemenite boy. She would probably be as heartbroken as Norman Podhoretz of *Commentary* says he would be if his daughter married a Negro, only he—and she—would try to make the best of it. It might be harder for Mr. Podhoretz to make the best of it than for the mother in Tel Aviv, but she would feel pangs. In Israel the chances are that intermarriage will erase differences in a way that Franz Boas hoped for in the United States. But North African, Yemenite, and Indian Jews do not belong to the "Negro race," as the term is used by the anthropologists. These Jews are not Negroid at all; they are dark Caucasoids, regardless of technical classification. There are probably no more than a few score Negroid Jews on either side of the Jordan; they would seem to have little future in Israel.

There is a fairly large batch of black Jews waiting in and near Ethiopia for the "return," without which they believe the Israelites cannot be redeemed by the Lord. They are the Falasha Jews, who are Negroid in appearance. Their hero is an Israeli, the son of the Chief Rabbi of Eritrea, who also served as rabbi for the Falasha Jews in Ethiopia. His real name is Colonel Hazi Oved, and he led the Druze battalions that sided with Israel in the War for Independence. He is also a scholar and has written on the Falasha with much sympathy and a good deal of romanticism. The first full-scale description of the Falasha Jews is probably to be found in James Bruce's *Travels to Discover the Source of the Nile*, published in Edinburgh in 1790, from which I quote the following:

> As we are about to take leave of the Jewish religion and government, in the line of Solomon, it is here the proper place where I should add what we have to say of the Falashas . . . who are reported to have come originally from Palestine. I did not spare my utmost pains in inquiring into the history of this curious people, and lived in friendship with several, esteemed the most knowing and learned among them, if any of them deserve to be so called; and I am persuaded, as far as they knew, they told me the truth.
>
> The account they gave of themselves, is that they came with Menilek from Jerusalem, so that they perfectly agree with the Abyssinians in the story of the Queen of Saba, who, they say, was a Jewess, and her nation Jews, before the time of Solomon; that she lived at Saba, or Azaba, the myrrh and frankincense country upon the Arabian Gulf. They say further, that she went to Jerusalem, under protection of Hiram, King of Tyre, whose daughter is said, in the XLVth psalm to have attended her hither, that she went not in ships, nor through Arabia, for fear of the Ishmaelites, but from Arabia round by Masuah and Suakem, and was escorted by the Shepherds, her own subjects to Jerusalem; and back again, making use of her own country vehicle, the camel; and that hers was a white one, of prodigious size, and exquisite beauty. . . . They say that when the trade of the Red Sea fell into the hands of strangers, the cities were abandoned, and the inhabitants relinquished the coast; that they were the inhabitants of these cities, by trade mostly brick and tile makers, potters, thatchers of houses, and such like mechanics . . . and, finding the low country of Dembea afforded materials for exercising these trades, they carried the articles of pottery in that province to a high degree of perfection, scarcely to be imagined.

> Being very industrious, these people multiplied exceedingly, and were very powerful at the time of the conversion to Christianity, or as they term it, the apostasy under Abreha and Atzbeha. At that time they declared a prince of the tribe of Judah, and of the race of Solomon and Menilek, was their sovereign. The name of the Prince was Phineas, who refused to abandon the religion of his forefathers, and from him their sovereigns are lineally descended; so that they still have a prince of the house of Judah, although the Abyssinians, by way of reproach, have called his family Bet Israel, intimating that they are rebels, and revolted from the family of Solomon and the tribe of Judah; and there is little doubt that some of the successors of Azarias adhered to their ancient faith also. . . .
>
> The Falasha are a people of Abyssinia, having a particular language of their own; a specimen of which I have also published. I do not mean to say of them, more than of the Galla, that this was any part of those nations who fled from Palestine on the invasion of Joshua. For they are now, and ever were, Jews, and have traditions of their own as to their origin. . . .

Another account of the Falasha Jews was written by the Reverend Henry A. Stern, himself a convert from Judaism, in his *Wanderings Among the Falashas in Abyssinia*, published in London in 1862:

> Claiming a lineal descent from Abraham, Isaac, and Jacob, the Falashas pride themselves on the fame of their progenitors, and the purity of the blood that circulates in their veins. Intermarriages with those of another tribe or creed are strictly interdicted, nay, even the visit to an unbeliever's house is a sin, and subjects the transgressor to the pennance of a thorough lustration and a complete change of dress before he can return to his own home.

An overenthusiastic and fabulous account of the Falashas and other African Jews is to be found in *Travels in North Africa* by Nahum Slousz.

Culturally and economically, the Falashas are Jews living in a broadly Semitic environment, speaking the language of the country they have resided in since the most ancient times. Although Amharic is their everyday tongue, as it is for most Ethiopians, their books, like those of the Ethiopians, are written in Geez. Hebrew, despite romantic claims by some partisan observers, is largely unknown to them except for prayers. There is no movement among Israelis or other bodies of world Jewry to bring the Falashas to the promised land. But the Falashas say that even if all 13 million other Jews were to come to Israel, the nation would not survive unless it also gathered in the missing tribe of black Jews from the mountain fastnessess of Ethiopia.

The black Jews do indeed present a problem to Israel. Jews are no different from other people in finding acceptance of 50,000 such different men, women, and children difficult. What is Israel to do? Israeli intellectuals think that the political leadership and the rabbinate are in a moral dilemma, and they are right. But an influx of thousands of black Jews into the already troubled nation, surrounded by more enemies than friends, will not help Israel. Still, if Israel is indeed the land of *all* Jews, as Israeli law proudly proclaims, then the moral imperative is clear.

I wish that I knew what the people of Israel ought to do. The dilemma is moral, financial, and religious. *Are* the Falashas Jews, after all? This question is for the scholars and rabbis to decide. Meanwhile, as in so many problems involving the attitudes of groups and nations and races, the hearts of men will ultimately decide.

Perhaps the Falasha problem, as I have described it, provides some comfort for Mr. Booker of the *Amsterdam News*. He can argue that Israel is no better than other nations when it comes to color prejudice, but I doubt that this discovery will benefit the disciples of the new *apartheid* in the United States. They have built up a fictional "history" of Africa that is truly pathetic. I wonder how they would react to the harsh facts of African life as they have been reported by historians. If Stokely Carmichael and his Africa-oriented comrades were to start probing deeply, they might even find an Arabic manuscript of the thirteenth century, written by Hadzi el-Eghwatti, which states that the people of what is now Ghana are of Jewish descent, even though they had been forced to adopt Islam as their religion. "They pray at the stated hours except on the day of djemat (Friday) which they do not observe as the sabbath. They possess great wealth. Their women appear in the marketplace veiled; and converse in

Hebrew among themselves, when they do not wish to be understood." This habit of conversing in a language not understood by the outsider is known even today in the so-called "ghettos" of our big cities.

But reason does not dispel hatred, nor can it conquer common prejudice. Man can find a thousand reasons for being unreasonable. It *may* help, therefore, to remember, when some uninformed Negro speaks of the great Negro kingdoms in Africa, that he may be referring only to the largely mythical Kingdom of the Hebrew Shepherds or to the great kingdom of the Black Jews of Africa in the tenth and eleventh centuries. I emphasize what Booker T. Washington, E. Franklin Frazier, and other Negro scholars and writers have emphasized: *The Negro in the United States is part of the American cultural tradition and owes as much to Africa as other Americans owe to India. English like all Indo-European languages derives from Sanskrit.*

This aspect of the confrontation is worth emphasis, for earlier in this book we cited the figure 3 percent for Negroes participating in the 1967 riots. But a Harris Poll taken before the riots had indicated that approximately 15 percent of the Negro population would participate in riots if they were to occur and that even more would support them. The black-power boys took note of these frightening statistics and evidently acted upon them. Perhaps it is because of the 1 million Negroes represented in those percentages that Floyd McKissick has called in his Black Manifesto for the establishment of a separate state or a series of black enclaves in the United States.

That result would truly be tragic. It is necessary to see African history and myths in proper perspective. I suggest that the American Negro does not need "new" history or a new language like Swahili; instead he should reread the Negro writers of his own country who knew that only through the American democratic process will he attain absolute dignity.

Roy Wilkins is certainly correct to call for the teaching of more Negro history in our schools—without intruding African patterns into the curriculum. When he asks that American history textbooks include material on Nat Turner and others, let us include it, but let it be the truth. Let us show both Negro and white children how fanaticism—and Turner was a fanatic—can lead to disaster.

Chapter Twenty
NEGRO ANTI-SEMITISM IN THE "GHETTO"

There are some Jews who deny that anti-Semitism exists in the black "ghettos." They rely on surveys in which Negroes who are asked outright if they dislike Jews mostly answer that they do not. But Negro folkways have shown and sophisticated Negro intellectuals have commented upon the fact that black men tend to give the kinds of answers expected of them in a given situation. The following quotation from James Baldwin is an example:

> I remember meeting no Negro in the years of my growing up, in my family or out of it, who would really ever trust a Jew, and few who did not, indeed, exhibit for them the blackest contempt. On the other hand, being utterly civil and pleasant to them, in most cases, *contriving to delude their employers into believing that far from harboring any dislike for Jews, they would rather work for a Jew than for anyone else.* (*Commentary*, Winter 1948.)

(A survey conducted by the American Jewish Committee in 1968 showed that 7 out of 10 Negroes preferred to work for Jews.)

Twenty years later Mr. Baldwin repeated almost the same message. He is *not* essentially an anti-Semite, nor are most responsible leaders or spokesmen for the Negroes. They are describing a real situation. The Jew's anguish is all the more acute, for he has contributed money and time to those whom he considers less fortunate than he.

By mid-1968 Jacques Torczyner, president of the Zionist Organization of America, had discovered "black anti-Semitism," and urged an about-face from the past when "the American Jewish community had played a leading role in the civil rights movement to aid the American Negro." He pointed to the "heavy sums" that had been contributed in this effort. At about the same time, too, *The National Observer* carried a front page story under the heading "Black Anti-Semitism." The story zeroed in on the bitter experiences of a number of "tiny merchants in chaotic Washington," following the post-Martin Luther King riots in the nation's capital. Here we are informed how two black hoodlums walked into a liquor store and demanded two half-pints of Scotch, after which they smashed the bottles on the sidewalk, returned, and said: "You're in deep trouble Jew man. Deep, deep trouble. You done sold us two bad bottles, man. They done explode. Now make good, Jew man." The proprietor "made good."

Later when two Jewish merchants were killed, the friends of one of them took an ad in the local paper which asked, "Where will tragedy strike next? Today, the inner city. Tomorrow, the residential areas, the suburbs. Today, Ben Brown [one of the murdered]. Tomorrow?" And "tomorrow" Charles Schweitzer [who placed the ad] was killed on the very day "the black-bordered advertisement appeared," as the *Observer* put it. The reporter, talking to a Jewish merchant, quotes him as follows: "The extreme liberals, including many of my own race here in town, are making us a scapegoat."

Perhaps just one more item is worth recording because it points to a bizarre use of terror, something even the harried "tiny merchants" had never been confronted with—not even in their memories of the dark days in the old country. Such memories seem to be always with them to remind that no matter how safe they feel, danger always lurks in their midst.

"They have to go," said a Black Panther captain, speaking from his headquarters at 780 Nostrand Avenue, Brooklyn. The "they" he referred to were the Jewish merchants on the once-busy shopping street. Many of them had been there for years. Some of them wept openly, while waiting for an opportunity to sell out and leave. Almost all of them had devised protective devices to make their shops burglar-proof and riot-proof. They had put up iron gates and metal grilles to protect their store fronts and doors. Suddenly the district was flooded with leaflets, originating in the Bronx, demanding the removal of these gates and other defenses. The fliers called the merchants "colonialists," charging them with putting on "steel enclosures" because they "don't trust us." Underscored was the legend: "The money they are spending for these iron gates should be put back into the community from which it is taken to improve ghetto conditions." This leaflet was signed by the Bronx Action Committee.

Now Jews are familiar with *chutzpah,* having invented the term which translates roughly as "colossal gall." But to be warned to remove safeguards, the easier to rob them, this they had never experienced in their lives.

However, at about the same time, The American Jewish Committee held another meeting to aid the Negro. It heard reports of progress in conducting an interracial camp in St. Louis, a camp retreat for high school students in Los Angeles, urban-suburban school operations in Philadelphia and on-the-job training for secretarial help in Los Angeles, Philadelphia and Dallas. It issued a 100-page report on police-community relations, all the time reassuring itself—and its officers (if not its membership)—that things were beginning to look bright on that bright April 2, 1968. Then Martin Luther King was assassinated and all hell broke loose, and today the amicable relations the Committee persists in pursuing only results in more insults and more demands for it (and similar "beneficiaries") to please mind their own business. For instance:

When the leaders of Reform Judaism met in Montreal a couple of years ago, they were addressed by the chairman of the Social Action Committee of the Union of American Hebrew Congregations (whose name I withhold out of charity) who told them to beware of "Jewish landlords and ghetto profiteers," adding, "In this whole area of social justice, we Jews cannot be part of the crowd. Here, too, we are the chosen people—chosen to bear the message, chosen to risk position and security, chosen to pioneer. . . . Here we must fulfill our destiny as gadfly to the conscience of mankind." Now this is abrasive talk, insolent to the point of inviting destruction. And he was soon answered:

"We want to make it crystal clear to you outsiders and you missionaries, the natives are on the move! Look out! Watch out! That backfire you hear might be your number has come up. Get out, stay out, stay off, shut up, get off our backs, or your relatives in the Middle East will find themselves giving benefits to raise money to help you get out from under the terrible weight of an outraged black community." This was signed by the Society for the Liberation of the Black Man in America.

Now it is hoped that neither the "black liberation society" nor the Hebrew "social action committee" speak for the Negroes or Jews of America. If they do, we are all doomed. Suddenly, late in 1968, it began to dawn even on the American Jewish Committee, or at least to its executive vice-president, Bertram H. Gold, that "there was increased feeling among Jews that Jewish organizations should withdraw from the civil rights struggle." At this time, too, Rabbi Bernard Weinberger, a member of the policy-making body of the anti-poverty program in New York City defined some truths when he pointed out that "unlike affluent Jews who are far removed from the confrontation with slum living, many Jews share slums with Negroes in New York, Los Angeles, Detroit, Baltimore, Chicago and elsewhere." His message: Spurn "deceptive liberal posturing." But this is harder to do than to ask. And sometimes even the most liberal of Negro leaders seems unable to understand it. Or, is it possible, he does?

In May, 1968, Bayard Rustin came out with a statement that was as timely as it was misdirected. Misdirected because it was issued in a report by the Anti-Defamation League of B'nai B'rith (ADL) whose members need not be reminded of it (or do they?). It would have been significant if it had appeared in, say, the *Amsterdam News*, or in another Negro journal. Said Rustin: "I am not one who goes about apologizing for or explaining away Negro anti-Semitism. It is here. It is dangerous. It must be rooted out. We cannot say it is somehow different or not really important. We cannot sweep it under the rug. What we can and had better do is to understand it if we are to deal with it. . . . Consider the question of the Jew in the ghetto. If you happen to be an uneducated, poorly trained Negro living by your wits selling numbers, selling dope, engaging in prostitution, then you only see four kinds of white people. One is the policeman, the second is the businessman, the third is the teacher and the fourth is the welfare worker. In many cities, three of those four are predominantly Jewish. Except for the policeman, the majority of the businessmen, the majority of the teachers, the majority of the welfare workers are Jewish. Here again is the love-hate syndrome."

Now it is Mr. Rustin's notion that where hate is concerned "those you love come first." This replaying of the old Oscar Wilde tune is, as they say, irrelevant. What is relevant is that it is Mr. Rustin's way of "explaining away" Negro anti-Semitism. Why childless prostitutes should worry about teachers, or why small numbers players should be concerned with businessmen, or dope pushers with welfare case workers Rustin does not make clear. Certainly these "professionals," in Harlem and elsewhere, are not the kind that bring their scrubbed-clean children to school, nor do they make application for relief. The facts are they belong to the criminal element of the community—and every knowledgeable observer can testify that there is very little ethnic or religious prejudice in the underworld. The dope pushers in the Mafia and the old crew of Murder, Inc., did not hate one another because of "racial" differences. If they had any differences it was over distribution of the loot. Besides, and this is crucial to our thesis: Anti-Semitism comes from the top down, from the intellectual, from the educated, if you will, not from the downtrodden and outcast, as Mr. Rustin implies. The editors of *Liberator* and the *Black Panther*, John Hatchett, LeRoi Jones, and the scores of others who have been pouring out their contempt at the Jew are among the best educated in the black communities and are not at all representatives of the pimp, the prostitute, the dope pusher and the numbers seller. In fact, it is an intellectual who urges these professions on the Negro community. Hate is built from the top and filters down to the mass by ideologues who have a very special and extremely dangerous axe to grind.

Speaking openly about the problem, even if only to excuse it, is still better than pretending that it does not exist. It does exist, and it is more dangerous than most people believe. More money, more scholarships, more advice, and more "social righteousness," as Bayard Rustin defines the spirit of Judaism, will not solve it.

Another moderate Negro, who claims to be a friend of the Jews, shows more disdain for the Jewish spirit than do their avowed enemies. Henry Lee Moon, editor of *The Crisis*, organ of the National Association for the Advancement of Colored People, tells us that:

> Anti-Semitism among Negroes is a minority phenomenon, unrepresentative of the total community. . . . Of the European immigrant groups of the 20th century only the Jews have given consistent organized support to the Civil Rights cause. . . . The Jews we are constantly reminded know how to handle money, how to make it, invest it, multiply it; they have a deep reverence for learning, are passionately devoted to education and achieve academic distinction; *and, above all, Jews stick together and help one another* (emphasis mine).

Professor Daniel Bell, who has done much original work on the subject, quotes from a research study done by a graduate student who found that the non-Jewish working classes or those "trapped on the bottom rungs of the ladder of vertical mobility . . . cannot understand how the Jews are able to rise into the middle classes, unless it is because *they stick together*" (emphasis mine).

It is obvious that Mr. Moon intended to compliment the Jews on the staff of the N.A.A.C.P. when he wrote the quoted passage. But I would rather face up to the naked anti-Semitism of Eddie Ellis; he is more honest, even if wrong.

In the three *Liberator* articles that so upset the Jewish community and troubled the consciences of Mr. Baldwin and Ossie Davis, Mr. Ellis argued that, when Jews proffer help to the Negro, they have only their own aggrandizement in mind. Acting together they use the Negro as the "stop-gap" for anti-Semitic feeling. "One commonly used method (by the Jews) has been, not only this infiltration of negro (sic) organizations, but in some instances they have supplied the money and have taken over the leadership of pre-dominantly negro organizations. A case in point is the N.A.A.C.P."

An editorial note preceding the second in a series of three articles says: "In part one of this series, we attempted to explain the reasons anti-semitism exists in the Afro-American community. The killing of Nelson Erby an unarmed Afro-American by Policeman Sheldon Liebowitz and his subsequent white-washing to a 'grand jury' has contributed to the intense anti-semitism feeling in the ghetto." At least the intensity of the feeling is acknowledged, even though the cause is put simplistically. But I doubt that Mr. Baldwin resigned because of remarks about Mr. Liebowitz or the N.A.A.C.P. It was the third article that prompted the resignations. In it Mr. Ellis really lets go; he identifies every slumlord with the Jew, and every Jew with predatory designs on the black community.

> Historically speaking, the Jews were among the first to recognize the fantastic potential contained within the spheres of "captive Black America." Upon this realization *they* immediately set the machinery in motion, to project themselves through philanthropic organizations, as the "true" friend of Black folks. It's little wonder then, scarcely two years after the abolition of slavery, money from Northern "white liberals" in many cases Jews began to find its way

> south in support of "negro" educational institutions. . . . Worse than any other, the Rosenwald Foundation, established by Julius Rosenwald, contributed in great measure to the almost absolute dominance of negro colleges and organizationists by Zionists. For example: The initial money used to establish the N.A.A.C.P. came directly from the Rosenwald Fund; from its earliest beginnings one of the main supporters of the Urban League was Julius Rosenwald.

Mr. Baldwin and Mr. Davis made their resignations known through the American Jewish Committee. Why is difficult to understand. Perhaps Mr. Ellis' entirely unwarranted indictment of Julius Rosenwald and his fund, which had been so instrumental in building schools in the South, in providing Negro education where it was lacking, and in furnishing scholarships and grants-in-aid to Negroes—Baldwin included—hurt most.

It was one instance in which the group around the *Liberator* could have been isolated from public opinion, black and white. Yet, except for the highly publicized resignations, nothing was done; today in the nation's Harlems the advocates of *apartheid* are *gaining* influence, as manifested in the Black Power Conference, CORE's Black Manifesto, the growth of the Black Panther organization into a full-fledged political party—with a presidential candidate in the 1968 election—and as the even more separatist Black Power Conference, held in Philadelphia during the summer of 1968, demonstrated. This conference was an amazing affair. Such "moderates" as Whitney Young and Manhattan Borough President Percy Sutton attended and whites were excluded. It has been reported that among the programs considered was a call for a plebiscite to determine the necessity for a separate Black Republic in the United States. The guiding spirit of this conference was Milton R. Henry of Pontiac, Michigan, first vice-president of the Republic of New Africa (RNA)—a nation that will be carved out of the states of Alabama, Mississippi, Louisiana, South Carolina and Georgia. This venture is being supported by a number of other Black organizations, including the Malcolm X Black Hand Society headed by W. C. Anas Luqman. Another supporter is Ron Karenga of *US*—"Anywhere we are *US* is!"—whose hatred for the United States is comparable to his contempt for the Christian religion. ("The Christian is our worst enemy . . . Jesus said, 'My blood will wash you white as snow.' Who wants to be white but 'sick' Negroes! or worse yet—washed that way by the blood of a dead Jew.") By September 1, 1967, Floyd McKissick was prepared to implement his Black Manifesto by setting up a completely black enclave somewhere in the United States. A century after the Civil War, more than a half-century after the founding of the N.A.A.C.P., a dozen years after the school-desegregation decision of the U.S. Supreme Court, and after almost all necessary civil-rights laws have been enacted, many Negroes are talking of a separate nation or areas. But while they talk of black *apartheid* in a sea of whites, their hatred and resentment are directed against the one minority that stands as a perpetual challenge to those who believe that one has to be born on the right side of the tracks to "make it" in the United States.

This confrontation has always been unfair. It has been largely invoked by certain Jewish intellectuals. The Lubavicher Rabbi, a Hasidic divine of our time, has said of such intellectuals that they are not so much alienated from society as committed to destructive self-sacrifice, by which I assume he means that they are ready to sacrifice all Jewry for their own individual ideals. Russia offers an example.

In the fifty years since the Bolshevik Revolution, during which leading Jewish intellectuals helped to place the yoke of dictatorship upon the Russian people, almost every decree limiting the rights of Jews and debasing their God ("the God of Israel is a fascist") has been left for Jews within the ruling Communist Party to promulgate. Openly anti-Semitic cartoons, the like of which *later* appeared in Streicher's *Der Stünmer*, appeared first in the Yiddish-language Communist press. Later, the justification for the acts of strategic aggression fomented by Israel's Arab neighbors was echoed by statements of American Jewish Communist leaders who are committed to totalitarianism. The Jewish intellectuals of the New Politics who voted approval of the anti-Israel resolution, were simply bearing testimony to the apparent need for self-destruction of the Godless Jews who, having abandoned the *kiddush hashem*—or sanctification of the Name in face of forced conversion or desecration of the faith—have adopted this way of demeaning their faithlessness by assuming an ultra-humanitarianism that is as inhuman as it is debasing. If this indictment sounds harsh it is because I am no saint, yet I know what Rabbi Aaron the Great meant when he said: "I wish I could love the greatest saint as the Lord loved the greatest rascal."

The continuing "dialogue" between Negro and Jew in this country has been carried on mainly in panel discussions, interviews, and articles in the Jewish press. James Baldwin, Bayard Rustin, James Farmer, Kenneth Clark, and occasionally Ralph Ellison have confronted representatives of such Jewish organizations as the American Jewish Committee and the American Jewish Congress in publications like *Commentary*, *Judaism*, *Congress Weekly*, *Congress Bi-Weekly*, and *Midstream*. With perhaps the single exception of a piece in *Ebony* written by a reform rabbi, the dialogue has not been carried on in the Negro press.

Jewish intellectuals have almost never heeded Charles Silberman's admonition to speak out and refute gross distortions of how the Jew succeeded in the United States. Men like Nathan Glazer have tried to set the record straight; others have tried to shift the emphasis by citing the Japanese or some other visibly oppressed minority that overcame indignities heaped upon it during wartime. But almost never have Jewish intellectuals who join in the dialogue been able to recognize their own self-deceptions.

Committed, as most of them are, to an optimistic and secular view of society, they insist on believing what is perhaps the greatest single untruth of our time: that all men *can* live in peace as brothers during the upward march of civilization that guarantees education and material plenty for all.

Although yearning is not to be dismissed, it should not be mistaken for reality. In the real world, if moral values are to be upheld, we cannot tolerate for a single moment such statements as that by Andrew Kopkind in *The New York Review of Books;* he claims that "morality, like politics, starts from the barrel of a gun."

Of course, Mr. Kopkind is an extreme case. He represents a revival of the nihilism that, in the last decades of the nineteenth century, produced Jewish boys to hand out leaflets calling for pogroms directed at their own parents. A young Bronx Jew who heads one of the splinter parties that support Mao Tse-tung was asked why he should support the man who had said, "Israel is the Formosa of the Middle East and should be driven into the sea." He replied, "That is a sacrifice I am prepared to make."

Some of the young men who head the most extreme leftist organizations in the United States, the Progressive Labor Party, for example, are not all Jewish, but at least one is the son of a Jew who has worked all his life as an executive of a Jewish professional organization. And today the Trotskyist movement, founded by Leon Trotsky (born Bronstein) is the single most abrasive and anti-Semitic movement in the nation; as ruthless and contemptuous of Israel and the Jews as are the American Nazis.

The responsible leaders and spokesmen for both sides of the Negro-Jewish confrontation have observed the degree of hatred for Jews within the Negro community. Some say that it has been exaggerated and that it has affected only a few Negroes.

It is a truism that rationales for anti-Semitism usually indicated more deep-seated hatred than that expended in the mere use of the expletive "dirty Jew." When Negro writers suggest that the Jewish merchant in the Negro slums is engaged in some kind of shoddy enterprise to cheat the poor, uneducated Negro when he is befuddled with liquor, to cheat him of the rent, and so forth, it seems a familiar refrain to most Jews. This unfair indictment of the Jewish merchant is old in the tradition of anti-Semitism. In an obscure chapter by the French socialist writer Charles Fourier of more than a hundred years ago, one can find some of the earliest references to the Jew as businessman, whom he equates with Judas Iscariot.

This is no place in which to elaborate on the lunatic economic theories of Mr. Fourier, but his message is accepted today not only by anti-Semitic Afro-Americans and "moderate" Negro leaders but also even by certain Jewish intellectuals and rabbis. The latter genuinely believe that no one acts purely out of malice, without some reason. If "irrational" behavior occurs, there must be a sound reason for it. Therefore, if Negroes distrust Jews in Harlem, it must be because Jewish landlords let garbage collect in the hallways and charge exorbitant rents, grocers overcharge for bread and milk, butchers place their "fat thumbs" on the scales, and liquor-store proprietors are determined to befuddle the minds of the poor in the black slums. Such is the myth.

The reality is somewhat different, as anyone who has visited some of the Negro neighborhoods in this country knows. I have visited a great many of them, and I have lived, played and worked with and for Negroes. It is true that certain retailers, not necessarily Jews, mark up prices to offset the mass pilferage that is common in the Harlems of the nation. Claude Brown, in *Manchild in the Promised Land*, tells how common it is for the Negro youngster to swipe anything from the "Jewstore" that is not nailed down—and sometimes even what *is* nailed down.

The war between the fatherless boys in the "ghettos" and the storekeepers, especially the Jewish storekeepers, never ends. I say "especially" because the Jewish candy-store owner, with his kindness to children and his total commitment to nonviolence does occasionally give a bit of candy to a Negro youngster and speaks kindly to him. The black boy almost always regards this kindness with suspicion, and often the same youngster will return when the proprietor is busy and take. Soon he is stealing a pair of roller skates. If there is no father at home to demand that he return the pilfered merchandise, if his mother only inquires if the skates are the proper size, the boy will grow up to loot. The Jewish merchant, imagining the boy's shame at being arrested, refuses to call the police, but decides to be more careful—or to mark up the price on this or that item to make up for the loss. This marking up does happen, although not on the scale that some amateur sociologists claim. Why does the Jew remain in business in areas where he is distrusted, robbed, and every once in a while burned out? The question must be answered, and I hope to answer it in the next chapter.

Chapter Twenty-one

THE MORE THINGS CHANGE

Let it be noted that nothing the Jew—intellectual or businessman—has done in the past has helped to lessen the tension visible in every facet of the Jewish-Negro confrontation. This fact has sometimes led to the circular argument that the Negro is anti-Semitic because he is anti white, and, because the merchant is usually Jewish, he hates the Jew more. The Jewish explanation is, of course, that Jews are disliked because they do not do enough for Negroes but that the dislike is unjust because Jews do more than other minority groups do. Negro intellectuals like Bayard Rustin then claim that, because Jews are committed by their faith to aid the poor, they will have to go on helping Negroes, regardless of the latter's response. Some Jewish intellectuals accept this argument, but it is a bit too much for the simple Jewish businessman. He wants only to be allowed to live in peace and to do business as best he can even in the most adverse circumstances. Here is how a bit of dialogue went a couple of years ago:

"You are Jews and you remain Jews only when you stand for social righteousness," Mr. Rustin said in a discussion with Charles E. Silberman and Rabbi Arthur J. Lelyveld, reported in *Congress Bi-Weekly*, May 23, 1966. No one present pointed out to Mr. Rustin that, if his statement were true, there would be few Jews in the world. A Jew is a Jew for many reasons and, if he behaves righteously, good. If not he is still a Jew.

Mr. Rustin went on, "If you are going to remain Jews only so long as Negroes remain nice, give it up." Give what up? Give up being a Jew? Give up helping the Negro? No one, not even Rabbi Lelyveld, who should know something about the tenets of his faith, thought to correct Mr. Rustin. We can understand Mr. Rustin better when he says, "We love the Jew and we hate him." Even then it is proper to inquire, why either one? Why does the Negro *love* the Jew, and why does he *hate* him? Mr. Rustin forged on into embarrassing irrelevance: "Come off it, Jews, daughters of Jews, brothers and sisters. Observing the law against mixing meat and milk, or even circumcision, doesn't make you Jews."

Here is a summary statement of the thirteen articles of Jewish faith as promulgated by the great philosopher and codifier of the Law, Moses Maimonides: belief in the existence of God; demand for recognition of His indivisibility, His eternity, and His absolute claim to unity (monotheism); acceptance of Moses as the greatest prophet; affirmation of the divinity of the Torah and the unalterability of the Law; reliance on God's providence, His just rewards, and His just punishments; belief in the future coming of the Messiah and in the ultimate resurrection of the dead. There is no "social" righteousness there, not even a hint of it. If Rustin is correct, Maimonides was no Jew.

Evoking the name of the martyred apostle of black nationalism, Malcolm X, Mr. Rustin remarked how he had told Malcolm, who had been able through faith to turn a prostitute into a self-respecting woman, that religion may be able to find cures for one or even ten but that "It is the ghetto, the absence of work, which will make ten dope addicts for every one that is cured." Once again the "ghetto" is blamed for generating dope addicts and prostitutes. Yet two distinguished Jews who must have had some acquaintance with the true ghettos of Europe, if only by hearsay, did not correct him for his simplistic economic determinism.

As Maurice Samuel has said: "Let us forget the slur on the ghetto. . . . The ghetto, by heavens, was not impotence; it was potency under terrible handicaps." Let us indeed forget the "slur" on the ghetto. Jewish intellectuals should speak out, as Mr. Silberman has said Jews should speak out, when Negroes, no matter how affable or how misinformed, say unwise or unjust things. They should reply to Mr. Rustin: The Jewish ghetto did not make for addiction, except perhaps to *pilpul*, over-indulgence in studying the Talmud. It produced Tevyah the Dairyman and the literature of Sholem Aleichem among a score of other distinguished writers and scholars.

Mr. Rustin can seek causes of delinquency and prostitution in the cluttered tenements of Harlem. He can attempt to refute Daniel Moynihan by arguing that jobs at fair pay will automatically solve the problem of the Negro family "in the same way the difficulties of the Irish family were cleared up." But his efforts to rebuild the Negro family will be unsuccessful for the reasons indicated earlier. The present point is that Jewish intellectuals who have had opportunities to engage in fruitful discussion with Negro leaders have failed in their responsibilities by not stating the truths they know for fear, as Mr. Silberman warned, of offending the Negro movement.

Rabbi Lelyveld did say "that the Negro has come to expect more of the Jew." He went on in an effort to explain the emerging "Jewish backlash," as he termed it in 1966: "When the leadership of a local battle to preserve *de facto* school desegregation is Jewish, or when a prominent Jewish businessman is part of the entourage of a Southern racist, the disappointment (among the Negroes) is greater and the hurt is deeper."

I suggest that the Rabbi's distress is a put-on. The phrase "prominent Jewish businessman" is there only to engage the sympathies of the socialist-oriented Mr. Rustin. But just as Rabbi Lelyveld was warming up to the issue of "*de facto* segregation," Negro leaders dropped it. They began to call for all-Negro schools, with no white, especially Jewish, teachers or principals.

As summer vacation approached in 1968, teachers in some Brooklyn schools found in their mail slots little notes in which they were asked not to return. Some of these invitations read, "Out of here you no good Jew bastards" or "Don't come back here in September if you know what's good for you." Sugar was dumped into the gas tanks of cars belonging to some teachers. The "justification" for such attacks was that the largely Negro school population was not receiving a proper education from the "Jew bastards."

Naturally, many of the teachers asked to be transferred. Neighborhood committees were set up to demand Negro principals and teachers, regardless of qualifications, to fill the places of the departing staff. As the fall term approached, some of the schools, where agitation was strongest and threats to the teachers most virulent, capitulated. So far black power has triumphed. *De facto* segregation has been forgotten, but in 1969 the new demand is for "separate but unequal" schools for the nation's "underadvantaged" children, that is, for Negroes. The Ocean Hill-Brownsville experiment in school decentralization in Brooklyn which led to a strike by New York's teachers that closed down most of the city's schools became so embroiled in charges of racism and anti-Semitism that a polarization set in that may well affect education of children in the so-called central cities throughout the nation.

Once again it is a Jewish intellectual who has provided the rationale for nightmare. Professor Harry Passow of Columbia University's Teachers College wrote:

> The district's (District of Columbia) sights should be set not on making schools equal, but on devising whatever means are required to enable every child to develop his potential and to get his chance. To live with this conception of equal opportunity, the community must be willing and the schools must be able to furnish unequal education. . . . Unequal education to promote equal opportunity may seem a radical proposal, but it is in fact a well established practice with respect to the physically and mentally handicapped which must now be broadened.

As I understand it, Dr. Passow believes that Negro children—although he calls them "poor children"—should be treated as we treat the mentally and physically handicapped. This belief can only increase bitterness and anger, for it is humiliating to the Negro who believes that "black is beautiful."

The dignity of the Negro has perhaps never been understood by the liberal. If I stress the word "liberal" in these pages it is because it is at him that so much resentment is directed in almost all segments of the old civil-rights movement and certainly among the newer leaders. As one Negro has put it, "the northern white liberal" has understood his people less than the southern segregationist has. In the long run the South's adjustment will probably be easier than that of the North. However much the Negro is looked down upon in the South, his reality is almost never denied. In the North, however, he is the "invisible man," the man no white man talks with, plays with, or plans with—except in the abstract. There is some truth implicit in Dick Gregory's quip, "In the South, they don't care how close I get so long as I don't get too big. In the North, they don't care how big I get so long as I don't get too close."

William Faulkner always wrote of the Negro as a human being capable of, and often successful at, transcending the hell of his environment. Because environments can be altered more rapidly than can the stubborn nature of man and because politicians are politicians and votes are votes, it is possible to envision greater improvement in the conditions of the Negro in the South sooner than in the urban North.

In the North massive handouts of city, state, and Federal money are designed to help alter the type of family structure the Negro knows; his dignity is attacked so that in the end he is compelled to shout, "let me be!" The "emancipation" of any minority is ultimately a problem in "auto-emancipation." More important, despite all the civil-rights laws that are on the books, the Negro faces the same difficulties as do free men everywhere. As Ralph Ellison has put it:

> There is also an American tradition which teaches one to deflect racial provocation and to master and contain pain. It is a tradition which abhors as obscene any trading on one's anguish for gain or sympathy; which springs not from a desire to deny the harshness of existence but from a will to deal with it as men at their best have always done. It takes fortitude to be a man and no less to be an artist.

How old must one be to assert one's essential manliness? Perhaps the greatest damage now being done to young Negroes—especially in urban areas—is his subjection to the so-called "integrated" school. The shame of being not quite able to compete with his white classmates, especially with the Jewish boys who gobble up high marks as other boys devour cakes and candy, may so frustrate him that he comes to regard all learning as alien.

Unless he is basketball or track material, the Negro boy in an academic high school tends to be more inhibited in his learning than if he were to attend school in a more congenial (homogeneous) atmosphere. Professor Eli Ginsberg has observed: "A Negro student who attends an interracial school in the north may encounter other psychological obstacles. His teachers are usually white. This fact alone may inhibit the quality of his performance. The Negro may be further inhibited by repeated failures to meet the competition of better prepared white students." He will certainly feel left out and left behind by the more motivated students, especially those for whom homework is a lark. The biggest flaw in the Supreme Court decision of 1954 was the assumption that contiguity with white pupils in classrooms would somehow improve the quality of the Negro child's education. What the court failed to recognize is that the dreams of the different groups that constitute the total American community are not all the same. Nor is it desirable that they should be. All that decency demands is that no opportunity be denied the Negro to pursue his dreams. The ultimate triumph over adversity and the city slums is up to him. Sidney Hook has written, "The ethics of democracy presupposes not an equality of sameness or identity, but an equality of differences."

It is this "equality of differences" that so many Jewish liberals have ignored. They have instead proclaimed an identity of suffering with the Negro that has only turned the black man's anger to revulsion. Never once in all the continuing debate between Jewish and Negro intellectuals have the former stated their own defenses. When attacked, they have agreed with their attackers and have asked for understanding of those Jews who do not see things as they do. On the single occasion that Norman Podhoretz, editor of *Commentary*, tried honorably to come to grips with the subject he made such a bathetic plea for understanding of his own difficulties that he made the reader long for the candor of Claude Brown. Mr. Podhoretz's article, entitled "My Negro Problem—and Ours," recounted his youth in Brooklyn during the 1930s and his trouble in reconciling two conflicting notions that seemed to conflict with the life about him: that all Jews are rich and that all Negroes are persecuted.

He knew only *poor* Jews, and the Negro boys from whom he fled in terror seemed to him to have all the privileges that any healthy boy might desire. They played hooky with impunity; they ate candy bars while Norman had to eat spinach and potatoes; they seemed absolutely unafraid of teacher, truant officer, and principal. He envied them their absolute sense of freedom and wanted to be friends with them, but they kept beating him up. He reminisces with genuine candor, and his experiences are common enough to seem neither bizarre nor shocking. Other boys with undeveloped muscles have had similar experiences growing up in the city's streets. But a sense of guilt at his own candor seems to haunt him. What will Mr. Baldwin think of his story? Or Kenneth Clark? Or Ralph Ellison? He then states the unbelievable "our" part of the "problem," and we run away embarrassed:

> Not so long ago, it used to be asked of white liberals, "Would you like your sister to marry (a Negro)?" When I was a boy and my sister was still unmarried, I would certainly have said no to that question. But now I am a man, my sister is already married, and I have daughters. If I were to be asked today whether I would like a daughter

> of mine "to marry one," I would have to answer: "No, I wouldn't *like* it at all. I would rail and rant and tear my hair. And then I hope I would have the courage to curse myself for raving and ranting, and to give her my blessing. How dare I withhold it at the behest of the child I once was and against the man I now have a duty to be?

In a letter to *Harper's* (May 1967) Mr. Podhoretz had taken exception to a statement by Mr. Ellison that "some of the *Commentary* writers" are among the "new apologists for segregation." Mr. Podhoretz called the comment a "calumnious falsehood." To which Mr. Ellison replied:

> What can I say to Mr. Podhoretz? He's so hair-triggered of tongue . . . so frequently up in somebody's face demanding apologies, creating sad feelings of needless abhorrence; putting down the down and buttering up the up. So what ever can I reply to Mr. Podhoretz when he says "not a single word has appeared (in *Commentary*) that any remotely responsible reader could characterize as an apology for segregation?" Either he thinks that the record will disappear, or that I will be too intimidated to use it. He should live so long.

Now this last sentence is a *Yiddishism* containing a kind of slur that is unworthy of Mr. Ellison. The claim that "some contributors" have been segregationists is not borne out by the facts, regardless of how inept some arguments may have been. He goes on to quote Nathan Glazer's article in *Commentary* (December 1964), "The New Challenge to Pluralism," and adds:

> For while the Jew had "always assumed that disadvantaged groups. . . . should advance without disturbing the group patterns of American life," Negro leaders believed that ethnic and religious subcommunities so "protected privilege or created inequality" that Negroes could not advance without modifying those patterns. Hence the Negro challenge and the Jewish resistance; . . . hence the crisis. And while I accepted with minor reservations his description of the *Jewish* side of the crisis, his apocalyptic framing of Negro

> demands bewildered me. I simply couldn't see the Negro for the Glazer. . . . And then slipping momentarily into black face [if Al Jolson could do it, why shouldn't a social scientist?] he allowed that Negroes saw "nothing in the Negro group whose preservation requires separate institutions [not even a chitt'ling supper?], residential concentration, or ban on intermarriage, or rather, the only thing that might justify such group solidarity is the political struggle itself. . . . What other groups see as a value, Negroes see as a strategy in the fight for equal rights. . . ."

These statements are offensive, for Dr. Glazer is among the more reasonable and least apocalyptic writers on the subject of minorities in the United States.

Mr. Ellison's resentment does show that writers in Jewish periodicals (including Daniel Moynihan) should steer clear of the Negro problem unless they are prepared for acrid controversy. Mr. Ellison quotes William Faulkner to Dr. Glazer:

> *I would say that the Negro doesn't want to mix with white people any more than white people want to mix with the Negro. (He) simply wants the right to decide not to mix . . . What he wants is mainly the chance himself to decide to be segregated . . . .*

"So spoke Anglo-Saxon Protestant Mississippi aristocrat, William Faulkner," Mr. Ellison adds approvingly. I doubt that he would have approved of the same remark had it been made by Mr. Podhoretz, Dr. Glazer, or any other Jewish intellectual. But he knew that Mr. Faulkner spoke from the heart about people and conditions he knew, whereas the Jewish intellectual speaks only from the mind, in sociological language that sets Mr. Ellison's teeth on edge.

Speaking of the White Anglo-Saxon Protestant, or WASP, the following quotation from Claude Brown's review of *The W.A.S.P.* by Julius Horwitz is revealing:

> The second principal character is sold to the reader as the White, Anglo-Saxon Protestant, but—to be euphemistic—it is a fraudulent transaction. Everything about the W.A.S.P. including his nigger-phobia, expressed in his brave, indefatigable championing of nigger rights, supports the proposition that the W.A.S.P. is a psychotic Jewish liberal.

What both Mr. Brown and Mr. Ellison distrust in the Jewish participation in Negro-white relations in the United States is the Jews' tendency to overintellectualize the problem. Mr. Brown would much prefer that Mr. Horwitz and the editors of *Commentary* would once in a while let themselves go and say "nigger" when they mean "nigger," instead of treating the black bum and the black artist as stereotypes in a Freedom March charade. Almost all Jewish writers are guilty of this mistake and almost all Jewish writers and critics have touched on Negro themes. Saul Bellow, Bernard Malamud, Norman Mailer, Mr. Podhoretz, Dr. Glazer, Mr. Silberman, the men who head the various committees to protect Jews from defamation, professional men, reviewers, commentators, lecturers—they have all insisted on speaking as experts. The sociologists are the most dismaying to Mr. Ellison, as well they might be. In an interview with some young Negro writers, he said about sociology:

> What is missing today is a corps of artists and intellectuals who would evaluate Negro American experience from the inside, and out of a broad knowledge of how people of other cultures live, deal with experience, and give significance to their experience. We do too little of this. Rather we depend upon outsiders—mainly sociologists—to interpret our lives. It doesn't seem to occur to us that our interpreters might well be not so much prejudiced as ignorant, insensitive, and arrogant. . . . But when we Negro Americans start "*writing*" [or protesting, he might have added] we lose the capacity for abstracting and enlarging life. Instead we ask, "How do we fit into the sociological terminology? Gunnar Myrdal said this experience means thus and so. . . ." Well, whenever I hear a Negro intellectual describing Negro life and personality with a catalogue of negative definitions, my first question is, how did you escape, is it that you were born exceptional and superior?

Mr. Ellison's statement is incontrovertible. But the sociologists do not let up. The more things change, the more they are the same. Now the sociologists are talking about separate but unequal. If they have to find rationalizations for anti-Semitic cartoons—like one in a S.N.C.C. handout that shows a hand, tattooed with the Star of David in which is inscribed a dollar sign, holding a rope from which dangle a black man and Gamal Abdel Nasser, while a sword labeled "liberation" is about to cut the hangman's rope—

they will find them. And the Negro will persist in asking, "Why do they insist on helping us; why don't they go away?"

Besides Mr. Podhoretz's article on *his* Negro problem we have had dozens of similar articles. Milton Himmelfarb tried to convince Negroes that Jews do not strike back because of the tradition of *broygez;* Arnold Rose (who assisted Dr. Myrdal with *The American Dilemma*) offers the strange thesis that the city environment provides the basis for anti-Semitism among Negroes; Harry Golden, on the contrary, tells us that "philo-Semitism," or love of the Jews, diminishes with distance from the cities and closeness to "the small towns and rural communities of the agricultural plains" of the South.

Mr. Golden is a journalist, and Dr. Rose is a sociologist. Sociologists have done more to poison group relations in the United States than have all the bigots of North and South. The solution may finally be to keep the sociologist from expressing himself on any but the most innocuous subjects.

Fundamentally the problem of the Negro in the United States is for the Negro to solve. If he resents taking responsibility for his own plight, he must be reminded that only he can make his life what he wants it to be. If he replies that he *cannot* do it, as Mr. Baldwin keeps repeating, we must show him examples of those similarly placed who *have* done it. If he asks for instant acceptance as equal by all whites, as Claude Brown asks, then we must say that matters of the heart cannot be manufactured. If he insists on intruding where he is not wanted, where community consensus cannot allow him to enter in peace, we must remind him that he is asking, not for freedom and equality, but for the abdication of both.

Is separatism a viable solution or a "copout" from society, from the necessity of looking to this country's promise for the future rather than to its dead past? We tend to look backward for parallels instead of seeking the creative potential available in the present. Politicians, especially liberal politicians like those in Detroit and New Haven, augur future trouble, by accepting sociological shibboleths as if they were inspired gospel. They are not. The politicians, often lacking any experience of slum neighborhoods, believe that a bribe here, a promise there, a minipark, an open hydrant, a store-front city hall, or orders to police to refrain from making arrests can help ensure civic peace. Instead they encourage incidents like the one in which youthful members of the moderate

N.A.A.C.P. tore up the Mayor's office, in Milwaukee in 1967. Following this outrage, the New York membership of the junior N.A.A.C.P. denounced Roy Wilkins as "the most paranoiac leader black people have." They demanded that the United Nations station observers in the United States for "systematic genocide committed . . . on the Indians and the enslavement, genocide and continued subjugation of black people." In the last days of 1969 they were to demand that the N.A.A.C.P. be turned into the N.A.A.B.P.—the National Association for the Advancement of Black People.

The bandying about of words like "genocide" and false analogies between the extermination of the European Jews by the Nazis and the Vietnam policies of the United States recall the equation by American friends of Joseph Stalin's extermination and imprisonment of millions with the arrest of the nine Scottsboro boys or the imprisonment of Tom Mooney. The sin of men like Ellison and Baldwin and Wilkins is that they know the falseness of this equation but keep quiet for fear of rocking the boat—or something.

Although Jewish liberals can find arguments for the new separatism, American Negroes will gain nothing from it but a retreat to nowhere. During the nineteenth century and the early part of the twentieth there were anti-Semites in Europe and the United States who thought it would be a good idea to solve the "Jewish problem" by sending the Jews to Palestine. These anti-Semites, who would not stoop to destroy the Jews or even to converting them, wished only to be rid of them and have therefore become known as the "get-rid-of-the-Jews-Zionists." We have also noted how Abraham Lincoln and Thomas Jefferson thought to help the Negro *out* in much the same way.

The Negro *can* look to the Jewish experience to enrich his understanding, but it must not be thrust upon him. He must seek, and *he* must find. Advising him of the delights that separatism will bring is hardly salutary. When Harry Schwartz of *The New York Times* editorial board noted virtues in "Black apartheid" that some Negroes had not yet discovered, he was intruding in a way that has so often exacerbated Negro-Jewish relations. He wrote:

> For some middle class Negroes the idea of a black nation carved out of the United States has a special attraction.

> It would require a new bureaucracy. It would have jobs for Cabinet members, ambassadors, generals, school supervisors. . . . In a black nation, Negro lawyers, doctors, engineers, architects and journalists would be protected from white competition.

It is possible that Mr. Schwartz was engaging in a wicked leg-pull, but I doubt it. I think that he meant every word, even though it might rankle among Negroes whom he taunted with notions of "black architects, engineers, doctors, lawyers and journalists" in a nation of their own, *as if* the United States could not and would not absorb every available and competent professional man in the land.

Floyd McKissick rejoices mistakenly at the money that is pouring in from white men to support his latest effort to remove Negroes from the Brooklyn slums and settle them "elsewhere" (at an undisclosed place) in the country. His fund-raising efforts are being enormously rewarded, though for all the wrong reasons. As one white man said on hearing of the plan, "Good, I'll start a movement to give a buck to get rid of the niggers, myself."

But the liberal, especially the Jewish liberal, has had much experience with the kind of intellectual strip tease performed by Mr. Schwartz. In politics it is known as the "trial balloon" or the "planted leak." In his assertion that his plan would appeal to "some middle class Negroes," Mr. Schwartz was making an assumption about black-white relations that even the social workers and television journalists have failed to make. He saw where an appeal for *power* would find its most dedicated adherents. The statistics suggest that it is the better-educated, not the least-educated, Negro who takes part in riots, that it is the young middle-class Negro who is apt to belong to militant organizations like the Black Panthers. Of the eighteen black men who were arrested in New York and charged with possession of an arsenal and with plotting to murder several moderate Negro leaders, at least four were professional men, members of the Negro middle class.

The demands have shifted from integration to *apartheid*. Some Negroes now cry that black is *better* than white and that Negroes should be granted more opportunity than the white man. This attitude bodes ill for the nation and most of all for the Jews.

Chapter Twenty-two

THE FUTURE

Let us repeat Hannah Arendt's warning: When the mobs are in the streets the Jews had better beware. So far they have been hit only "in the pocketbook," as one young Negro said, referring to the burning and looting of stores during riots. But they have also suffered other casualties; for example, a seventy-year-old Jew was mugged and beaten by a group of young blacks in the Bronx. Other cities report similar incidents. This particular example is cited because it suggests the abdication of elementary moral principles throughout the nation. A police officer—a Negro—saw the attack and came to the aid of the victim; in his zeal he shot and killed one of the young muggers, whereupon a hue and cry went up for his arrest for "murder." Black-power groups were demanding his conviction, and the local white press referred to the old Jew as "a Caucasian fellow" without naming him. A new refinement was thus added to the arsenal of racism.

Nobody had a word for the victim of the beating. His name has been forgotten. The fourteen-year-old who was shot by the detective has become a martyr, referred to in the press as "the boy" or the Negro "youth."

It can be foreseen that such incidents will multiply in the future. And the mob is the new constituency of the politicians who rule our cities. A week after the second case of rape of a young girl in New York's Central Park while one of her male companions was almost stomped to death, Mayor John V. Lindsay spoke on television. A listener might have thought that he was speaking of a city in never-never land rather than of strike-torn and strife-torn New York.

But the white liberal believes that the rioter is only a youth who wants what the rest of us have. As for violence, he recalls the Molly McGuires of a century ago, the draft riots in New York during the Civil War, and so on. To the liberal as to the bigot, all Negroes not only look alike; *they are alike*. The liberal, for all his pity at the "plight" of the Negro, has really never gone out of his way to live with him, work with him, play with him, or fight with him. In a word, he is afraid of him. It is that fear probably more than anything else that raised Ralph Ellison's hackles in his exchange with Podhoretz in *Harper's*.

If the intellectuals engaged in the Negro-Jewish dialogue, which seems to serve as a prototype for a national black-white dialogue, cannot face each other without recrimination and insults, what can be expected from the antagonists in the ranks? When the "armed Bohemians" finish exhorting the masses and the "bloody-minded professors" finish indoctrinating their classes, there will remain only the confrontation of mobs in the streets. Yet it is premature to cry "pogroms," as the *Jewish Daily Forward* referred to the Watts riots, or to retreat into total separation of the races. Despite the hatred engendered by racism, it is still possible to hope that the best men on both sides will come to their senses.

Caution is called for and a different kind of separation. What is necessary is a separation of the Jewish intellectuals, with their special periodicals and organizations, from Negro attempts at self-assertion. There are enough Negro psychologists, writers, critics, artists, activists, moderates, conservatives, and ordinary hard-working people to lead the way to the kind of relative freedom that most of us enjoy in this country. The problem is for Negroes to solve.

Kenneth Clark once stated that he found it arrogant of certain Jewish rabbinical groups to undertake a study of the "Negro problem." It is all the more arrogant for Jewish writers to continue to approach the nation's racial problems as if they *as Jews* had a special expertise.

The sons of Hollywood tycoons who play God to the "underprivileged" in Watts have little enough talent to justify their presuming to serve as literary and spiritual guides to the untutored and unwashed. Seeds are planted that will surely blossom into a kind of Jew-devouring Mau Mauism that is too horrible to contemplate. We can see that kind of reaction in Claude Brown's review of Julius Horwitz's book, discussed in Chapter 21. We can see it in the open rebellion of the "moderate" youth battalions of the N.A.A.C.P. against their aging leaders, who want to temporize with the problem. The Jewish leaders of that organization have been subjected to so much hostility, from within the ranks and from outside, that one wonders why they remain. Certainly the Negro community has enough legal talent, for instance, to employ top counsel from its own ranks. When Dr. Clark says that he now sees the white liberal more in terms of the adjective than the substantive, he is simply describing an uncomfortable truth. He too, it will be recalled, thought that the golden era of color-blindness in social relations had been ushered in by the U.S. Supreme Court in 1954.

Color is here to stay for a while. Let us recognize it and act accordingly, so that it will no longer be necessary to assume that the Negro elevator operator is a bosom brother or that the Negro professor is the same as the Negro elevator operator. Each wants to be seen as what he is. Negroes are a differentiated group, with artists, writers, athletes, singers, doctors, lawyers, professors, and hoodlums; in that respect they are like every other ethnic group.

If the Negro wants to achieve *his* dreams *his* way, who are we to deny him? But it will not do to make believe that he has achieved or desires to achieve the same things that most Jewish intellectuals desire for their children. It is possible, of course, that a proper view of heaven requires a special vision. If the Negro thinks his path is blocked by obstacles put in his way, then others should not clutter up his landscape.

This point can best be illustrated by a little story told by Bayard Rustin, who wants the black-white dialogue to continue:

> The roots of all this disturbance were built in. I'll give you an example. A Negro girl down South is working in a Snick headquarters where she has been told about "participatory democracy." Get this scene—*and I saw this myself.* The girl is struggling to prepare a mimeographed press release which has to be out by 3 o'clock. She can't type or spell. A white girl comes in, looks at it in horror, says, "Move over," and knocks it off in three minutes. That happens over and over. (New York *Post,* June 23, 1967)

The liberal's expectation of instant acceptance and instant improvement is one cause of riots. Another is the black militants' cry that "We won't wait" or "We have waited too long." The farther removed from the slavery experience we come the more it seems to haunt white men. Negroes like Stokely Carmichael who invoke it probably never even knew anyone who had really experienced slavery in his father's or even his grandfather's generations. E. Franklin Frazier recalls a girl who had never been farther south than northern Illinois yet who told him the most harrowing tales of conditions in the South. When he asked her what part of the South she was from, she replied that she had never been there but that she had read all about it in the Chicago press. Irving Howe, the Jewish intellectual, embarrasses Ralph Ellison, the Negro novelist and critic; Julius Horwitz angers Negro author Claude Brown; some Jewish rabbis incur the mild contempt of Negro psychologist Kenneth Clark; and James Baldwin repeats that the antagonism between Negro and Jew is a "perpetual" one. Finally, there is the comic side of the confrontation. Dick Gregory says, "Every Jew in America over thirty years old knows another Jew that hates Negroes, and if we hate Jews, that's just even, baby."

I do not question Mr. Gregory's "statistics." I question only his humor. Yet for years a special audience has laughed at his jokes for fear that otherwise it will be considered intolerant. He does sometimes exhibit a wry humor, but his penchant for dubious statistics too often comes to the fore. He was quoted after the Newark riots as having said that they were a good thing, in that they probably prevented fifteen riots elsewhere. After this bit of wisdom had been duly reported on television, he commented on the Detroit riots that they had probably prevented *fifty* other riots. A truly funny man might have answered that, after Detroit, fifty others would have been superfluous.

But there is uglier "humor" in the situation. Suddenly cartoons appeared showing Jews in the usual anti-Semitic spangles of dollar signs and six-pointed stars. Worse, young Jewish philosophers, sociologists, and educators helped to feed the new anti-Semitism by seeking parallels between the mystique of totalitarianism and the personality of Moshe Dayan.

We are back in Germany in the mid-1930s, when young leftist Nazis—the "roast beef" Nazis, all brown on the outside and red within—painted *Judenrein* on the windows of Aryan stores. Now rioters avoid stores sporting "Soul Brother" signs. In Germany only stores were attacked at first ("the Jews have plenty, which they have stolen from us, and we are only taking back what is rightfully ours"). A little later came the single beatings and killings, but not so many as to cause great dismay even among the Berlin Jews, who were sure the trouble would all blow over. Then the big killings came.

In the 1960s H. Rap Brown, out on bail on charges of inciting to riot, described the Plainfield, New Jersey, ambuscade of the summer of 1967 as exemplary, for there armed rioters had so demoralized the civil administration that it had ultimately retreated from its sworn duty to maintain civil peace. It backed away because of a misguided application of the principle of civil liberty, the same misapplication that ultimately helped to kill democracy in Germany.

A similar threat confronts the Jews of the United States. All the laws on the statute books will not protect the Jews, who constitute no more than 3 percent of the total population (less than one-quarter the Negro population), unless state and city administrations enforce them.

In Germany the rich Jews symbolized everything the poor German coveted, yet Jews were high in the parties of social democracy and communism. From the latter ranks, the little man first heard the cry, "Hang the capitalist Jews." The ordinary German, whom Hitler courted and whom he assured that he would know how to impose Marxism "better than the Communists," was frightfully confused.

In the same way the American Negro sees the Jew as defender of his faith to a degree that he is compelled to cry out in warning, "Don't whip yourself." The American Negro is no German yet! He has not *yet* been too much indoctrinated with the hatred of Jews so typical of European politics, but he is learning fast.

So in the *Liberator* we read reams of copy dedicated to the proposition that the Jew has nothing but ulterior motives in coming to the aid of the Negro. The six-day war between Israel and its Arab neighbors is reported in terms distorted by racial hatred, and Israel is denounced so bitterly that Syrian Baathist fanatics seem bloodbrothers to the black-nationalist fanatics. In an address before a national Arab Student Assembly, Stokely Carmichael pledged black support for the Arabs against Israel. In Ocean Hill and in other predominantly Negro school districts in New York leaflets were widely distributed by black extremists declaring: "It is impossible for the Middle East murderers of colored people to possibly bring to this important task (educating black children) the insight, the concern, the exposing of the truth that is a must if the years of brainwashing and self-hatred that has been taught to our black children by those bloodsucking exploiters and murderers is to be overcome."

The crescendo mounts. Old euphemisms are discarded. There is now need for the higher defamation. "Slumlord" for Jewish landlord is not needed anymore; the new, more volatile, attacks on the Jew as the "imperialist Zionist bandit" is fetched forth—and here the ideology for which Jews have given so much, perhaps too much—the spirit of "socialism"—is invoked with which to whip them into further humiliation. At the close of the year 1969, Eldridge Cleaver called (from Algeria) for *jihad* against all Zionists and their supporters. But in Harlem (at just about this time) a self-appointed black spokesman for the community roared his defiance at the demands of more moderate Negro leaders for a State Office Building by declaring it was all "a conspiracy to let the New York Jewish Establishment continue to exploit blacks and Puerto Ricans." Suddenly—and without warning—the Puerto Rican community was invited to join in the chorus of hate against the Jewish "conspiracy."

But Jews have insisted that they have something special to teach Negroes, that Jewish suffering must mean something to Negroes. In fact it means nothing to the masses of Negroes. Hearing about it "increases the Negro's rage," as James Baldwin put it.

No liberal Jew, and most remain liberals even when their success might logically turn them into conservatives, will ever understand this point. They will protest, argue, cite statistics, defend themselves, but they will never understand. The reason is their optimistic philosophy; they are the last remaining disciples of world brotherhood, internationalism, and color-blindness in a world riven by racism, nationalism, tribalism, ambition, greed, and power. They believe these iron passions can be obliterated with the magic word: education.

In the New York public schools, where Jewish children constitute less than 10 percent of the total enrollment, Jews hold almost all the principalships and a majority of the teaching jobs. Negroes, whose children represent a much larger proportion of the school population, are woefully underrepresented in the education profession. Negro children are on the average two to three years behind their classmates in reading and other elementary skills. To the Negro parent unschooled in educational subtleties, the correlation seems simple: Jewish teachers do not educate *his* children properly. His anger is fanned by the words of the extremist, who raves that the Jews, with malice aforethought and exotic designs too devious to explore are deliberately miseducating the black child.

Meanwhile, lengthy studies have shown that the Watts rioters were not primarily the illiterate members of the community; most had high-school diplomas, and many were not among the most "underprivileged." Certainly the community was not a slum, as that term is generally understood; Watts enjoys spacious lawns, one- and two-family houses, and wide streets and highways.

Militant anti-white groups have mushroomed as a result of the riots and of the common excuse that "institutions," rather than antisocial inclinations, are responsible for their behavior. The Black Panthers and paramilitary movements like *US* are now well known. *The New York Times* saw fit to print the first sympathetic report on the Panthers by Sol Stern, an editor of *Ramparts.* Mr. Stern reports that one of the leading Panthers told him how they go about ragging the police, to the point at which they can march up to the cop during a coffee break, and then "they shoot him down—voom, voom—with a 12-gauge shotgun." According to Mr. Stern, "Like most revolutionaries, the leaders of the Black Panthers do not come from the bottom of the economic ladder." Some of them have even attended college. Ron Karenga, a leader of *US,* believes that "We live in a political society and what is important is power not morality." These gun-carrying black "revolutionaries" are not to be dismissed lightly. Mr. Stern warns: "What matters is that there are a thousand black people in the ghetto thinking privately what any Panther says out loud." That *is* frightening.

The Panthers have now emerged as the party of separatism in America, with defense ministers, information ministers, etc., as if they were already a nation within a nation. Many young Negroes on the college campuses seem to give them allegiance as do some of the nation's best amateur athletes. This is probably due to two factors: the assassination of Martin Luther King, Jr., and the release from prison of Eldridge Cleaver, author of *Soul on Ice.* Mr. Cleaver has been able to supplant some of the crude anti-cop agitation by a more militant—if that is possible—demand for self-determination. For the Panthers, as for numerous other activists, the violent death of Dr. King "proved" the illogic of nonviolence. Following the assassination of Senator Robert F. Kennedy, and the collapse of Senator Eugene McCarthy's campaign for the Democratic presidential nomination, some frustrated white radicals have flocked to the Panthers.

Dexter Knox, a black writer, ridicules their support of the Black

Panthers by an attack on David McReynolds, a white leader of New Left politics:

> Remember it was YOU who gave notoriety and license to these betrayers to run wild in the community. It was YOU who let them jet around the country. It was YOU who got them book writing—if you can call it that!—jobs, television press conferences and coverages, radio speeches, magazine articles, ad infinitum. It is YOU who force these bad elements onto the black communities throughout the country as OUR 'leaders'. . . .

The cult of violence grows, often encouraged by so-called "liberal" publications, by huge foundations, and by ladies of vast wealth. Frantz Fanon is the prophet. "Violence is a cleansing force," he has written. "It frees the native from his inferiority complex and from his despair and inaction. It makes him fearless and restores his self-respect."

Andrew Kopkind wrote in *The New York Review of Books* that "morality, like politics, starts from the barrel of a gun." Apparently what is powerful is right, and what fails is wrong. Hitler once said, "Right is what is good for the movement."

The first formal proclamation of the Cheka, forerunner of all the Soviet secret police organizations, reads, "All is permitted for us."

Nikolai Lenin wrote, "We repudiate all morality, outside of class morality." Today that morality too is also being cast out, even as a propaganda slogan. The new apostles and disciples of the cult of violence can do without morality; violence is now preached as an end in itself.

Where it will all lead I do not know. But one thing is certain: The American Jew who abhors violence has everything to lose in a violent confrontation; the Negro extremist has everything to gain. But in any confrontation based on anti-Semitism and hatred the Negro must surely lose his freedom ultimately, as the German did when he donned the Nazi uniform.

Understanding and reasoned argument will not avail the Jew in the Negro's struggle, He has already humiliated himself, apologizing for his temerity, his wealth, his education, his affluence, his liberalism, his commitment, his martyrdom. Now it is time for him to depart from the field with honor. A total withdrawal from the "ghetto," with goods, possessions, wife, and children, is necessary. He must leave his business, but he must not permit himself to be kicked out. He must make one final effort to demonstrate his good will to Negroes in an area where he is most competent—and most vulnerable: his shop or the property he owns but does not live in.

Jewish storekeepers in Harlem, Watts, or anywhere between—encamped as if in enemy territory—can expect forays into their business establishments when the next riots break out. They should *now* offer to train—immediately—local Negro men to take over their stores at fair prices and after adequate periods of preparation.

Jewish businessmen in the Negro communities of the land have suffered enough abuse and misunderstanding from their own intellectuals and even their own rabbis, to say nothing of Negro intellectuals. They have lived to see their daughters, educated to be schoolteachers, weep as demagogues take over classrooms to teach children false history. Adding bitterness to their tears is the fact that until the day before yesterday they had been paid-up members of the Student Non-Violent Coordinating Committee. The fathers understand that their daughters betrayed themselves.

The Jew must go with dignity. He must not give way to violence. He does not need the advice from secular Jewish spokesmen, men whom newspapers, radio, and television insist on identifying as spokesmen for all Jews in the land. No racial or national group in this country is a monolithic bloc, for which we may all be grateful. Experienced Jewish merchants can easily go elsewhere if necessary, thus allowing the growth of "indigenous" Negro enterprises. At first the Negro businessman will require help and serious ambition to achieve; he should not be distracted by raucous cries of "socialism" before he has so much as tasted the fruit of the free-enterprise system.

Perhaps the stores will have to be "manned" by women at first. Among Jews such situations have often enhanced family solidarity in the long run. The Negro family needs cohesion more than it needs freedom rides and freedom marches. The missing husband—statisticians tell us that millions of Negro men between the ages of thirty and forty are simply not included in the national census because they have disappeared—may return when he can acquire new status as storekeeper or merchant.

Furthermore, when he can build his shop into a chain—or a factory—he can follow the path of other minority groups and hire his kinfolk and neighbors. As studies have demonstrated, at first Jewish dress manufacturers usually employed relatives for the choice jobs in their shops; the same was true of Italian construction bosses and of Chinese laundry and restaurant proprietors. Negroes have not only the right but also the duty to do the same.

A slogan like "Operation Tradesman" or "Business Transfer and Training Center" can be adopted; Negro banks and established middle-class enterprises can help underwrite the first trying years of the experiment. To help deflect, perhaps even to avoid, a future bloody confrontation between Jews and Negroes, Jewish shopkeepers must make the changeover as painless and mutually profitable as possible. Here the Negro cannot have it both ways. He cannot demand the business and later refuse to run it unless someone subsidizes it, or because he refuses to be "a goddam storekeeper," as one of them put it.

Such efforts to engage Negro interest in business as the program announced by the Interracial Council for Business Opportunity should help in some small measure to overcome resistance in this area. But too much cannot be expected because of the council's "mix" and its top-heavy approach. Organized by the Urban League and the American Jewish Congress and funded by the Federal government and several foundations, it has already received $750,000 to continue work in New York, Newark, Los Angeles, and Washington, D.C. Its past performance has been spotty and largely unproductive, mainly because it has relied on the paternalistic professionalism of organizations not *in* business, so that the element of risk is made to seem negligible. The sponsoring organizations have not shown much awareness of or interest in *why* Negroes shun business. Such projects usually find their financial commitments consumed by administrative costs and by planning that is never translated into reality.

Dr. Alvin Poussaint, the Negro psychiatrist, thinks that "central to the entrepreneurial spirit is assertiveness, self-confidence and the willingness to risk failure in an innovative venture. A castrated human is not likely to be inclined in any of these ways." Castrated or not, the Negro will simply have to go it alone with whatever help *he* requests. And he will have to take his risks. The Jewish businessman can do more *with* him—not *for* him—than all the foundations in the land. The disinterested foundation may help to underwrite the risk, but a good insurance policy is better. And the best insurance for each member of our confrontation is to part friends, not anguished, envious, contemptuous enemies. The Jewish merchant in the so-called "ghetto" can provide an out for himself and can bolster the kind of pride the Negro needs as no one else can.

The Jewish shopkeeper has little to lose and much to gain, up to and including his own life. In the end, a hundred Negro-owned stores that formerly belonged to Jews can do more to improve so-called "intergroup relations" than can a hundred interracial and interfaith meetings conducted in an atmosphere of desperation. Only then will it be possible to achieve a more natural kind of intercourse.

For his own soul, for his own future, for his own peace of mind, the Jew must separate himself at every level from participation in Negro affairs and must persist only on terms of individual friendship wherever possible.

Friendship is a healthy relationship, for it makes possible the communication of truth that is unacceptable from enemies. The Jews in the N.A.A.C.P., in the Urban League, in CORE, should resign. Let them not wait for the fire next time. Once is enough; twice is too much; the third time may be a holocaust.

But, more important than this partial, or group disengagement, it is time for the nation to declare a 2-year moratorium on the "Negro Problem." Let this time be put to good use to consolidate past gains—and assess past failures. In any case, things won't dry up, or stand still entirely, but the heat of discussion must be lowered, tempers deflated—and "solutions" put on ice for a reasonable cooling-off period.